PYTHON PROGRAMMING FOR BEGINNERS

THE ULTIMATE STEP-BY-STEP GUIDE TO PYTHON PROGRAMMING WITH TIPS AND TRICKS TO MASTER PROGRAMMING QUICKLY WITH PRACTICAL AND EASY EXAMPLES

GEORGE RICHARD

TABLE OF CONTENT

CHAPTER 1

INTRODUCTION

PYTHON SOFTWARE

Python is a high-level, object-oriented, interpreted programming language with dynamic semantics. Combined with dynamic typing and dynamic linking, its high-level data structures make it very attractive for Rapid Application Creation and for use as a scripting or glue language for connecting existing components. Python's easy to learn syntax emphasizes readability and therefore reduces the cost of maintaining the program. Python supports packages and modules which promote modularity of the program and reuse of code. The Python interpreter and the comprehensive standard library are available free of charge in source or binary form for all major computer Platforms, which can be distributed freely.

Programmers also fall in love with Python because of the improved productivity it provides. The edit-test-debug cycle is incredibly fast because there is no compilation step. Python debugging programs are simple: a bug or bad input will never cause a segmentation fault. Instead, it raises an exception when the interpreter discovers an error. If the program fails to capture the exception, the interpreter will print a stack trace. A source-level debugger allows you to inspect local and global variables, evaluate arbitrary expressions, set breakpoints, walk through the code one line at a time, etc.

The debugger itself is written in Python, referring to the introspective strength of Python. On the other hand, sometimes, the easiest way to debug a program is to add a few print statements to the source: the quick process of edit-test-debug makes this simple approach very successful.

IDLE

IDLE is an integrated development environment that comes with Python. It's a program that lets you type in and run your programs. There are other Python IDEs, but I would suggest sticking with IDLE for now as it is easy to use. On your computer, you can find IDLE in the Python folder.

It starts up in the shell when you first start IDLE, which is an interactive window where you can type in Python code and view the output in the same window. I frequently use the shell instead of my calculator or try out small pieces of code. But you'll want to open a new window most of the time and type the program in there.

Note At least on Windows, if you click on a Python file on your desktop, the program will be run by your system, but not the code, which is probably not what you'd like. Instead, if you right-click on the file, an option called Edit with Idle should be available. To edit an existing Python file, either do that or start IDLE, and open the file via the File menu.

CHAPTER 2

DOWNLOAD AND INSTALL PYTHON

GET PYTHON

The most popular and up-to-date source code, binaries, documentation, news, etc., are available on Python's official website: http:/www.python.org

Python documentation is available for download at www.python.org/doc/. The documentation can be found in the formats HTML, PDF, and PostScript.

PYTHON INSTALLATION

Distribution of Python is available for a wide range of platforms. You just need to download your platform 's appropriate binary code, and install Python.

Python - Environment

If your platform's binary code is not available, you will need a C compiler to manually compile the source code. Compiling the source code gives you more flexibility in choosing the features you require when installing your system.

Here's a quick overview of how to install Python on different platforms:

UNIX AND INSTALLATION OF LINUX

Easy steps to install Python on Unix / Linux machine are provided here.

- Go to http:/www.python.org/download a web browser.

- Follow the link to download the available zipped Unix / Linux source code.

- Download files and extract them.

- Module/Install file editing if you want to customize any choices.

- Run a Script/Configure

- Let it happen

- Make installation

This installs Python at the default /usr / local / bin location and its libraries at /usr/ local /lib/ pythonXX where Python version is XX.

WINDOWS INSTALLATION

Here are the steps to get Python installed on the Windows machine.

- •Open your web browser and visit http:/www.python.org/download/
- Follow the python-XYZ.msi file link to the Windows installer where XYZ is the edition you need to download.
- Windows program must support Microsoft Installer 2.0 to use this installer python-XYZ.msi. Save the installer file to your machine then run it to find out if MSI supports your computer.
- Run the file. This brings up the very simple to use Python update wizard. Just accept the default settings, wait for the installation to finish, and you're done.

INSTALLING MACINTOSH

Recent Macs come with Python installed but may have been out of date for several years. For guidelines on getting the current version along with additional tools to support development on the Mac, see http:/www.python.org/download/mac/. MacPython is available to older Mac OS's before Mac OS X 10.3 (launched in 2003).

It is maintained by Jack Jansen, and you can access the entire documentation on his website- http://www.cwi.nl/~jack/macpython.html. Total configuration info can be found for

INSTALLING ON MAC OS

Fixing PATH

Programs, as well as other executable files, can be in many directories. Therefore, operating systems have a search path that lists the directories the OS is looking for executables.

The path is stored in a variable environment, which is a maintained operating system named string. This variable contains the command shell and other programs with the information.

The path variable is called a PATH in Unix or Path in Windows (Unix is case-sensitive; Windows is not).

The Installer handles the details of the path in Mac OS. You must include the Python directory to your path to invoke the Python Interpreter from any particular directory.

UNIX / LINUX PATH SET

In Unix, add the Python directory to the path for a specific session:

- Type setenv PATH "$PATH:/usr / local / bin / python" in csh shell, and press Enter.
- In the Linux bash shell: type ATH="$PATH:/usr / local / bin / python "export and press Enter.
- Type PATH="$PATH:/usr / local / bin / python "and press Enter in sh or ksh.

Note: The path to the Python directory is /usr / local / bin / python

In Windows setting path,

Adding the Python directory to the path for a given Windows session:

Type path % path%; C:\Python at the command prompt, and press Enter.

Observe: C:\Python is the Python directory path

Variables in the Python environment

Here are essential variables of the environment that Python can recognize:

Variable	Description
PYTHONPATH	It has a similar function to that of PATH. This

	variable tells the Python interpreter where to find the imported module files inside a program. It should include the directory of the Python source library, and the directories containing the source code of Python. Often, PYTHONPATH is preset by the Python installer.
PYTHONSTARTUP	It contains the path of a Python source code containing an initialization file. Every time you start the interpreter, it is executed. It is named in Unix as .pythonrc.py, and it contains commands loading utilities or modifying PYTHONPATH.
PYTHONCASEOK	In Windows, it is used to instruct Python to find a case-insensitive match in an import statement. Set any value to that variable to activate it.
PYTHONHOME	It is a search path of an alternative node. Usually, it's embedded in the PYTHONPATHr PYTHONSTARTUP

	directories to make switching libraries easier.

INTERACTING WITH PYTHON VIA IDE

An Integrated Development Environment is an application that combines more or less all the functionality you've seen up to now. Usually, IDEs have REPL capabilities as well as an editor that allows you to create and change code for execution by uploading it to the interpreter.

You will also come across interesting features like:

- • Highlighting the syntax: IDEs often color the different syntax elements in the code to make reading easier.
- • Code-completion: Some Integrated Development Environment can complete partially typed pieces of code (such as function names) for you — a great time-saving feature and convenience.
- • Context-sensitive help: Advanced IDEs may display Python documentation related information, or even suggested fixes for common types of code errors.
- • Debugging: A debugger helps you to step-by-step execute code and inspect program data as you go. This is invaluable when you try to determine

why a program is behaving inappropriately, as is inevitably the case.

IDLE

Most Python installations have a basic IDE, which is called IDLE. The name supposedly stands for an Integrated Production and Learning Environment, but Eric Idle is known as one member of the Monty Python troupe, which hardly seems like a coincidence.

The way IDLE runs differ from one operating system to another.

IDLE IN WINDOW STARTUP

Go to the Start menu, then select All or All Apps programs. A program icon should be labeled as IDLE (Python 3.x 32-bit), or something similar. This will be slightly different for Win 7, 8, and 10. The IDLE icon may appear in a folder of the program group called Python 3.x. You can find the IDLE program icon from the Start menu by using the Windows search facility and typing it in IDLE.

Click on the IDLE icon to start.

HOW TO START IDLE ON MAC OS

Check to open spotlight. One of several ways to do this is to type in Cmd+Space. Type terminal in the search box, and press Enter.

Type idle3 in the terminal window, and press Enter.

IDLE starts on Linux

IDLE is available for distribution with Python 3 but may not have been installed by default. Open a Terminal window to find out if it is. This varies depending on the Linux distribution, but by using the desktop search function and searching for a terminal, you should be able to find it. Type idle3 in the terminal window, and press Enter.

If there is an error saying command not found or anything to that effect, then obviously IDLE is not installed, so you need to install it.

The way apps are installed also varies from one Linux distribution to the next. For instance, the command to install IDLE with Ubuntu Linux is Sudo apt-get install idle3. Many Linux distributions have application managers based on GUI, which you can also use to install apps.

To install IDLE, follow whatever procedure is appropriate for your delivery. Type idle3 into a terminal window, then press Enter to run it. Your installation process may also have set up a system icon to start IDLE somewhere on your desktop.

HOW TO USE IDLE

Once you have successfully started IDLE, you should see a window titled Python 3.x.x Shell, where 3.x.x corresponds to your Python version:

The >>> prompt ought to look familiar. You can type REPL commands, just like when a console window starts the interpreter. Bearing in mind the universe qi, show Hello, World! Once again:

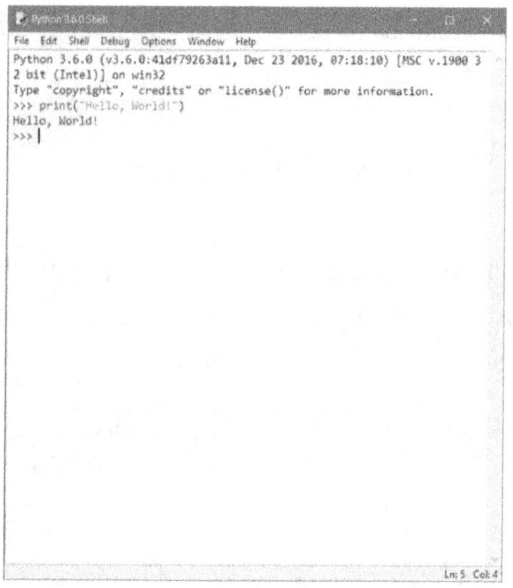

The interpreter has the same behavior as when you ran it directly from the console. To make things more readable, the IDLE interface adds the advantage of displaying different syntactic elements in different colors.

It also offers support sensitive to the context. For instance, if you type print(without typing any of the arguments into the print function or closing parenthesis, flyover text should appear specifying the print) (function's usage data.

Another feature IDLE provides is the Reminder statement:

- • If you have typed in multiple statements, you can recall them in Windows or Linux using Alt+P and Alt+N.

- • • Alt+P cycles backward through statements previously executed; Alt+N cycles forward;
- • • Once a statement is recalled, editing keys on the keyboard can be used to edit it and then implement it again. In Mac OS, the respective commands are Cmd+P and Cmd+N.

You can build and run script files in IDLE too. Pick New File from the Shell window menu. That would open a window for more editing. Type the code you wish to execute:

Select Save or File from the menu in that window and save the file to disk. Then select Run module to run. The output will appear in the Shell interpreter window again:

Okay, that's probably enough. Hello, World! The qi will be safe in the universe.

You can switch back and forth once both windows are open, editing the code in one window, running it, and displaying its output in the other. IDLE offers a basic forum for developing Python in this way.

Although somewhat basic, it supports quite a bit of additional features, including code completion, code formatting, and a debugger. For more information, please see the IDLE report.

Thonny

Thonny is a free Python IDE, developed and maintained by the University of Tartu, Estonia, Institute of Computer Science. Specifically aimed at beginners in Python, the GUI is plain and uncluttered, as well as easy to understand and quickly become familiar with.

Like IDLE, Thonny supports interaction with REPL and the editing and execution of script files:

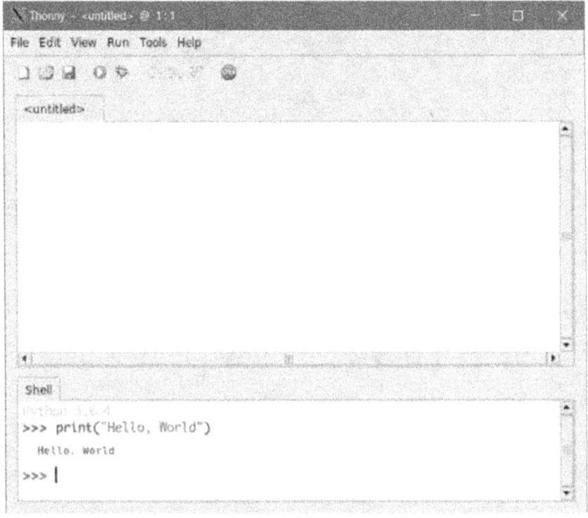

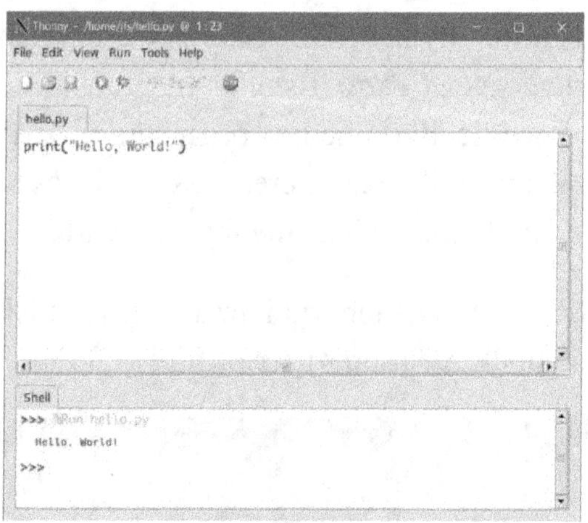

In addition to providing a step-by-step debugger, Thonny performs syntax highlighting and code completion. One function that is especially helpful to those studying Python is that when you step through the code, the debugger shows values in expressions as they are evaluated;

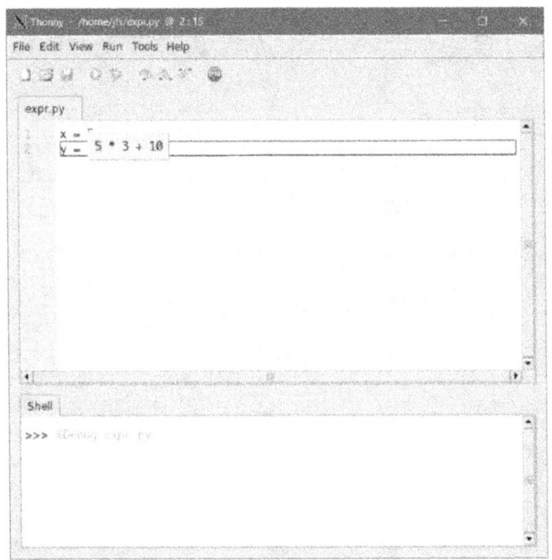

Thonny is particularly easy to get started because it comes with built-in Python 3.6. So you just have to do one install, and you're ready to go!

Windows, Mac OS, and Linux versions are available. The Thonny website has instructions for downloading and installing.

IDLE and Thonny certainly aren't the only games to go. There are lots of other IDEs available for editing and developing Python code. For additional suggestions, see our Guide to Python IDEs and Code Editors.

CHAPTER 3

THE PYTHON LIST

The data structure most frequently used in Python is List. Python list is a container that holds an ordered sequence of objects, like an array. The object may be anything from a string to a number, or any available type of data.

A list may be both homogeneous and heterogeneous, too. It means that we can only store integers or strings, or both depending on the need. Next, each item rests at someplace in the list (i.e., index). The index can later be used to locate a given item. The first index starts at zero, and the next is one, and so on.

Since in Python programming, the use of lists is frequent, you need to learn that well. This tutorial will help you learn- how to create and use a list in Python programs. It will also teach you about the various operations of the list, such as slicing, searching, adding, and deleting elements.

Please ensure you have an IDE to write and run Python code before proceeding. Python comes with an IDE known as IDLE, by default. This is pretty simple, but decent enough for learning.

If you want an actual IDE experience, though, then read the post below and pick one.

Best Python IDE

Create a Python List

In Python, there are several ways to form a list. Let's start with the most effective one.

Operator Subscript

The () square brackets represent Python's subscript operator. It does not require a lookup symbol or a call to a method. Hence, creating a list in Python turns out to be the quickest way.

You may define elements within () which are divided by commas.

Create a Python list using the subscript operator

Syntax

L1 = [] # An empty list

L2 = [a1, a2,...] # With elements

The list can take any number of elements, and each may belong to another type (a number or a string, etc.).

blank list

L1 = []

list of integers

L2 = [10, 20, 30]

List of heterogenous data types

L3 = [1, "Hello", 3.4]

Constructor List ()

Python includes a built-in list) (constructor method a.k.a.

It either accepts a sequence or tuple as the argument and converts to a Python list.

To create a list without any element, let's start with an example.

Create Python list using list()

Syntax

theList = list([n1, n2, ...] or [n1, n2, [x1, x2, ...]])

We can provide the list) (function with a standard or nested sequence as the input argument. Let's construct An empty list first.

theList = list() #empty list

len(list)

0

Note-The len () returns the size of the list.

Below are some more examples showing the process Python list ().

>>> theList = list([1,2])

>>> theList

[1, 2]

>>> theList = list([1, 2, [1.1, 2.2]])

>>> theList

[1, 2, [1.1, 2.2]]

>>> len(theList)

3

List comprehension

Python supports a concept called "List Comprehension" which helps to construct lists in a completely natural and easy way.

A list comprehension has the following syntax

#Syntax - How to use List Comprehension

theList = [expression(iter) for iter in oldList if filter(iter)]

It has square brackets which group an expression followed by a clause for-in and zero or more if statements. The outcome is always to be a list.

Let's look at a basic example first.

>>> theList = [iter for iter in range(5)]

>>> print(theList)

[0, 1, 2, 3, 4]

Isn't a simple list easy to start with.

Here is a more complicated example of List Comprehension leading to the creation of list.

>>> listofCountries = ["India","America","England","Germany","Brazil","Vietnam"]

>>> firstLetters = [country[0] for country in listofCountries]

>>> print(firstLetters)

['I', 'A', 'E', 'G', 'B', 'V']

Comprehension of the list also allows an if statement to add only members who meet a particular requirement to the list:

>>> print ([x+y for x in 'get' for y in 'set'])

['gs', 'ge', 'gt', 'es', 'ee', 'et', 'ts', 'te', 'tt']

Let's see now how the if clause works with understanding the list.

```
>>> print ([x+y for x in 'get' for y in 'set' if x != 't' and y != 'e' ])
```

['gs', 'gt', 'es', 'et']

Another complicated example is to create a list with a List Comprehension syntax containing the odd months.

```
>>> months = ['jan', 'feb', 'mar', 'apr', 'may', 'jun', 'jul', 'aug', 'sep', 'oct', 'nov', 'dec']
>>> oddMonths = [iter for index, iter in enumerate(months) if (index%2 == 0)]
>>> oddMonths
```

['jan', 'mar', 'may', 'jul', 'sep', 'nov']

Read List Comprehension to see more details.

Multidimensional list creation

By specifying an initial value for each element you can create a sequence with predefined size.

```
>>> init_list = [0]*3
>>> print(init_list)
```

[0, 0, 0]

You can compile a two-dimensional list with the concept above.

two_dim_list = [[0]*3] *3

The above statement works but instead of creating separate objects, Python will create the references as sublists only.

```
>>> two_dim_list = [ [0]*3 ] *3

>>> print(two_dim_list)

[[0, 0, 0], [0, 0, 0], [0, 0, 0]]

>>> two_dim_list[0][2] = 1

>>> print(two_dim_list)

[[0, 0, 1], [0, 0, 1], [0, 0, 1]]
```

We changed the third item value in the first row, but it also affected the same column in other rows.

So, to get around the above issue, you have to use list comprehensions.

```
>>> two_dim_list = [[0]*3 for i in range(3)]

>>> print(two_dim_list)

[[0, 0, 0], [0, 0, 0], [0, 0, 0]]

>>> two_dim_list[0][2] = 1
```

```
>>> print(two_dim_list)
[[0, 0, 1], [0, 0, 0], [0, 0, 0]]
```

Extending a list

Python makes it possible to re-size the lists in many ways. Just adding two or more of them can do that.

```
>>> L1 = ['a', 'b']
>>> L2 = [1, 2]
>>> L3 = ['Learn', 'Python']
>>> L1 + L2 + L3
['a', 'b', 1, 2, 'Learn', 'Python']
```

Extend List () Example

Alternatively, you can use the extend() method to join lists.

```
>>> L1 = ['a', 'b']
>>> L2 = ['c', 'd']
>>> L1.extend(L2)
>>> print(L1)
['a', 'b', 'c', 'd']
```

Example of List Append()

Next, by calling the append() (method you can add a value to a list. See Example below.

>>> L1 = ['x', 'y']

>>> L1.append(['a', 'b'])

>>> L1

['x', 'y', ['a', 'b']]

DIFFERENCE BETWEEN TWO PYTHON LISTS

Indexing list

There are many ways you can access or index the elements in a Python list. We listed some here.

Index operator

The simplest way to access an element from the list is by using the index operator]). Since the list has zero as its first index, there will be indices from 0 to 9 in a list of size ten.

Any attempt to access an item beyond that range would lead to an IndexError. The index is an integer every time. The use of any other value type will result in TypeError.

Also note that the nested indexing will be preceded by Nested list. Let's take only a few examples.

```python
vowels = ['a','e','i','o','u']

consonants = ['b', 'c', 'd', 'f', 'g', 'h', 'j', 'k', 'l', 'm', 'n', 'p', 'q',
'r', 's', 't', 'v', 'w', 'x', 'y', 'z']


#Accessing list elements using the index operator

print(vowels[0])

print(vowels[2])

print(vowels[4])


#Testing exception if the index is of float type

try:

 vowels[1.0]

except Exception as ex:

 print("Note:", ex)


#Accessing elements from the nested list

alphabets = [vowels, consonants]

print(alphabets[0][2])
```

print(alphabets[1][2])

a

i

u

Note: indices for the list must be integer or slice, not float

i

d

Reverse indexing

Python allows reverse indexing (Negative) for the type of sequence data. So you need to set the index using the minus(-) sign for the Python list to index in the opposite order.

The indexing of the list with "-1" returns the last element of the list, -2 the last and so forth.

vowels = ['a','e','i','o','u']

print(vowels[-1])

print(vowels[-3])

Output

u

i

List slicing

Python comes with a magic slice operator returning the portion of a sequence. It operates on objects with different types of data such as strings, tuples, and it functions the same on a Python list.

Syntax Cutting

It's got a magical syntax, which is the following.

#The Python slicing operator syntax

[start(optional):stop(optional):step(optional)]

Say size => Total no. of elements in the list.

Start (x) -> It is the point (xth list index) where the slicing begins. (0 =< x < size, By default included in the slice output)

Stop (y) -> It is the point (y-1 list index) where the slicing ends. (0 < y <= size, The element at the yth index doesn't appear in the slice output)

Step (s) -> It is the counter by which the index gets incremented to return the next element. The default counter is 1.

Let's take an integers list below.

>>> theList = [1, 2, 3, 4, 5, 6, 7, 8]

We'll be checking different slice operations on this list in the examples to follow. You need to know that we can not only use a slice operator to slice but also to reverse and copy a chart.

Slicing lists

Here are some examples of the use of indexes to slice a chart.

Returns the three elements from the list, i.e.[3, 4, 5]

>>> theList[2:5]

[3, 4, 5]

The first index starts at 0, since the Python list follows the zero-based index rule.

You can see, therefore, that we passed '2' as the starting index because it contains the value '3,' which is included by default in the slice.

And passing '5' as the end index meant to ensure that elements up to the 4th index can be included in the slice output.

Print slice as [3, 5], do not change index first or last

>>> theList[2:5:2]

[3, 5]

In this example, to exclude the median value, i.e., '4' from the slice output, we increased the phase counter by '2'

Slice from the third index to last item second

As for the stop argument, you can use a negative value. It means the traverse starts from the index at the rear.

A negative stop value like '-1' would have the same meaning as "length minus one."

>>> theList[2:-1]

[3, 4, 5, 6, 7]

Get the slice from beginning to index second

If you do not mention the "start" point in slicing, then it indicates beginning to slice from the 0th index.

>>> theList[:2]

[1, 2]

Slice from index two to finish

If the stop value is missing while slicing a list, then it indicates slicing to the end of the list. It saves us from passing list length as the ending index.

>>> theList[2:]

[3, 4, 5, 6, 7, 8]

Reverse a list

To achieve this, it is effortless to use a special slice syntax (:-1). But please remember that this is more memory-intensive than a reversal on the spot.

Use Slice Operator to reverse a list

Here, a shallow copy of the Python list is generated, which needs long enough space to hold the entire list.

>>> theList[::-1]

[8, 7, 6, 5, 4, 3, 2, 1]

You need to pause and consider here, why is the '-1' after the second colon? It intends to increase the index by -1 each time, and guides traversing in the reverse direction.

Switch a list back but leave values at odd indices

You can use the concept learned in the previous example here.

>>> theList[::-2]

[8, 6, 4, 2]

By setting the iteration to '-2' we can skip every second member.

Python Add two Elements List

Copy a list below

Let's see how Slice Operator produces a list duplicate.

Create a slim copy of the complete list

```
>>> id(theList)

55530056

>>> id(theList[::])

55463496
```

Since all the indices are optional, we can leave them out. A new copy of the sequence will be created.

Copy of the list which contains all other elements

```
>>> theList[::2]

[1, 3, 5, 7]
```

Python Iterate List

Python delivers a traditional for-in loop for list iteration. The "for" statement makes the processing of the elements of a list one by one super simple.

```
for element in list:

print(element)
```

If both the index and the element are to be used, then call the enumerate () function.

for index, element in enumerate(theList):

print(index, element)

If you want the index only, then name the methods range() and len().

for index in range(len(list)):

print(index)

The list elements support the iterator protocol. To intentionally create an iterator, call the **built-in iter function**.

it = iter(theList)

element = it.next() # fetch first value

element = it.next() # fetch second value

Check out example below.

theList = ['Python', 'C', 'C++', 'Java', 'CSharp']

for language in theList:

 print("I like", language)

<u>Output</u>

I like Python

I like C

I like C++

I like Java

I like CSharp

Append elements to a list

Unlike the string or tuple, the Python list is a mutable entity, so that it is possible to change the values at each index.

The assignment operator (=) may be used to update an element or a range of items.

Operator of Assignments

theList = ['Python', 'C', 'C++', 'Java', 'CSharp']

theList[4] = 'Angular'

print(theList)

theList[1:4] = ['Ruby', 'TypeScript', 'JavaScript']

print(theList)

<u>Output</u>

['Python', 'C', 'C++', 'Java', 'Angular']
37

['Python', 'Ruby', 'TypeScript', 'JavaScript', 'Angular']

You can also refer the add / extend section of the list to update the list.

Method Lists insert ()

You can also move one element by calling the insert () method at the destination location.

theList = [55, 66]

theList.insert(0,33)

print(theList)

Output

[33, 55, 66]

You can use the Slice assignment to insert multiple items.

theList = [55, 66]

theList[2:2] = [77, 88]

print(theList)

Output

[55, 66, 77, 88]

Delete items from list

Python provides various ways of adding or deleting elements from a list. Some of these include the following:

User Del

The keyword 'del' can be used to delete one or more things from a list. Additionally, the entire list item can be removed as well.

```python
vowels = ['a','e','i','o','u']

# remove one item

del vowels[2]

# Result: ['a', 'e', 'o', 'u']

print(vowels)

# remove multiple items

del vowels[1:3]


# Result: ['a', 'u']

print(vowels)

# remove the entire list

del vowels

# NameError: List not defined

print(vowels)
```

Remove () and POP ()

To remove an item from the desired index you can call remove() method to delete the given element or pop() method.

In the absence of the index value the pop) (method deletes and returns the last element. This is how lists can be defined as stacks (i.e., FILO-First in, last out model).

vowels = ['a','e','i','o','u']

vowels.remove('a')

Result: ['e', 'i', 'o', 'u']

print(vowels)

Result: 'i'

print(vowels.pop(1))

Result: ['e', 'o', 'u']

print(vowels)

Result: 'u'

print(vowels.pop())

Result: ['e', 'o']

print(vowels)

```python
vowels.clear()

# Result: []

print(vowels)
```

Operator Slice

Last but not least, by assigning an empty list with a slice of its elements, you can remove items too.

```python
vowels = ['a','e','i','o','u']
```

```python
vowels[2:3] = []

print(vowels)
```

```python
vowels[2:5] = []

print(vowels)
```

Output

```
['a', 'e', 'o', 'u']
```

```
['a', 'e']
```

Check elements in a list

Some of the standard search methods include:

In owner

Python 'in' operator can be used to search whether there is an element in the list.

if the value in the list:

    print("list contains", value)

Index List ()

Using the index () method of the Python list, you can find out the position of the first corresponding item.

loc = the List.index(value)

The index method performs a linear search, and breaks after the first matching item is located. If the search ends without result, it will throw an exception to Value Error.

try:

    loc = the List.index(value)

except ValueError:

    loc = -1 # no match

if you want to retrieve the index, for all matching objects, call index () in a loop by passing two arguments – the value and the starting index.

loc = -1

try:

    while 1:

```python
        loc = theList.index(value, loc+1)

    print("match at", loc)

except ValueError:

  pass
```

Here is a better version of the code given above. We wrapped the search logic inside a function in this and named it from a loop.

Example

```python
theList = ['a','e','i','o','u']


def matchall(theList, value, pos=0):

  loc = pos - 1

  try:

    loc = theList.index(value, loc+1)

    yield loc

  except ValueError:

    pass


value = 'i'
```

for loc in matchall(theList, value):

    print("match at", loc+1, "position.")

<u>Output</u>
match at 3 positions.

Python list supports two Min(List) and Max(List) methods. You can call them to find out which element carries the minimum or maximum value accordingly.

>>> theList = [1, 2, 33, 3, 4]

>>> low = min(theList)

>>> low

1

>>> high = max(theList)

>>> high

33

Sorting a Python List

List method sort)

Python list implements the sort) (method for ordering its elements in place (both in ascending and in descending order).

theList.sort()

Please note that in-place sorting algorithms are more efficient, as temporary variables (such as a new list) are not needed to hold the result.

By default the sort) (function performs ascending sequence sorting.

theList = ['a','e','i','o','u']

theList.sort()

print(theList)

['a', 'e', 'i', 'o', 'u']

If you wish to sort in descending order , please see the example below.

theList = ['a','e','i','o','u']

theList.sort(reverse=True)

print(theList)

['u', 'o', 'i', 'e', 'a']

Built-in sorted () function   The built-in sorted) (function can be used to return a copy of the list with its elements ordered.

newList = sorted(theList)

By default, it also sorts in an ascending manner.

```python
theList = ['a','e','i','o','u']

newList = sorted(theList)

print("Original list:", theList, "Memory addr:", id(theList))

print("Copy of the list:", newList, "Memory addr:",id(newList))
```

Original list: ['a', 'e', 'i', 'o', 'u'] Memory addr: 55543176

Copy of the list: ['a', 'e', 'i', 'o', 'u'] Memory addr: 11259528

You can turn on the "reverse" flag to "True" for enabling the descending order.

```python
theList = ['a','e','i','o','u']

newList = sorted(theList, reverse=True)

print("Original list:", theList, "Memory addr:", id(theList))

print("Copy of the list:", newList, "Memory addr:",id(newList))
```

Original list: ['a', 'e', 'i', 'o', 'u'] Memory addr: 56195784

Copy of the list: ['u', 'o', 'i', 'e', 'a'] Memory addr:7327368

PYTHON LIST METHODS

List Methods	Description
append()	A new element is added to the end of the list.
extend()	It extends a list by adding elements of a different list.
insert()	To the desired index, it injects a new element.
remove()	The required item is removed from the list.
pop()	It both removes an item from the given position and returns it.
clear()	All components of a list are flushed out.
index()	It returns the index of an entity that first matches.
count()	It returns the total no of passed elements as an argument.
sort()	It ascends in order of the elements of a list.
reverse()	This inverts the elements in a list order.

copy()	It executes and returns a shallow copy of the list.

PYTHON LIST BUILD IN FUNCTIONS

Functions.	Description.
All().	It returns True if the list includes or is blank with items with a True value.
Any().	If any of the members have a True value, then True returns as well.
Enumerate().	It returns a tuple with an index of all the list elements and their value.
Len().	The return value is list size.
List().	It converts all objects iterable and returns as a list.
Max().	The member whose highest interest is
Min().	The member whose minimum value is

Sorted().	It returns a copy of the list sorted.
Sum().	The return value is the sum of all elements in a list.

CHAPTER 4

DICTIONARIES IN PYTHON

Python provides another type of composite data called a dictionary, similar to a list, as it is an object collection.

Here's what you'll learn in this tutorial: You'll cover Python 's basic dictionary features and learn how to access and manage dictionary data. Once you've completed this tutorial, you should have a good sense of when and how to use a dictionary as the right type of data.

Dictionaries and lists share the below characteristics:

• They are both mutable.

• They are both dynamic. They can grow and shrink whenever appropriate.

• Both may be nesting. A list could contain a different one. A dictionary can contain a different dictionary. A dictionary can include a list, too, and vice versa.

Dictionaries differ mainly in the way elements are accessed from the lists:

• The list elements are accessed via indexing by their location in the list.

• Keys are used to access dictionary elements.

To describe a Dictionary

Dictionaries are the implementation by Python of a data structure, more generally known as an associative array. A dictionary is a collection of pairs with key-value. Each key-value pair maps the key to its corresponding value. You can define a dictionary by enclosing a list of key-value pairs in curly braces({})separated by a comma.

A colon (:) sets each key apart from its associated value:

```
d = {

    <key>: <value>,

    <key>: <value>,

    .

    .

    .

    <key>: <value>

}
```

The following describes a dictionary that maps a place to the name of the Major League Baseball team in question:

```
>>>

>>> MLB_team = {

...     'Colorado' : 'Rockies',

...     'Boston'   : 'Red Sox',

...     'Minnesota': 'Twins',

...     'Milwaukee': 'Brewers',

...     'Seattle'  : 'Mariners'... }
```

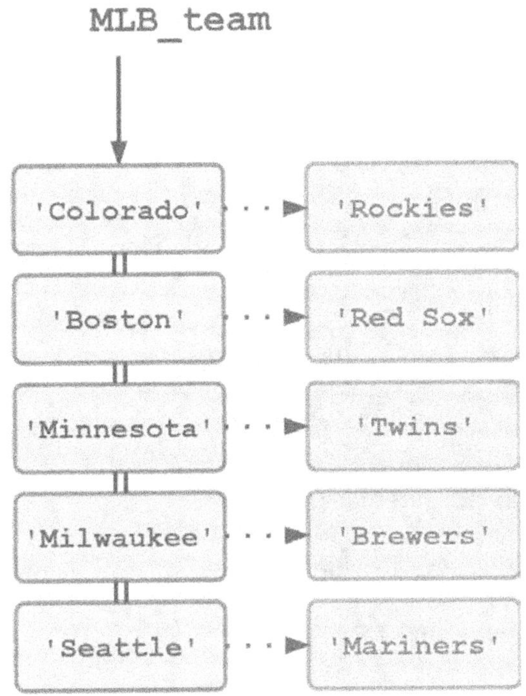

Dictionary Mapping Location to MLB Team

You can also build dictionaries using the built-in dict)
(function. The dict) (argument should be a series of pairs
of key-value. With this a list of tuples works fine:

```
d = dict([

    (<key>, <value>),

    (<key>, <value),

    .
```

    .

    .

    (<key>, <value>)

])

MLB_team can then also be defined this way:

>>>

>>> MLB_team = dict([

...   ('Colorado', 'Rockies'),

...   ('Boston', 'Red Sox'),

...   ('Minnesota', 'Twins'),

...   ('Milwaukee', 'Brewers'),

...   ('Seattle', 'Mariners')

...])

If the key values are simple strings, the keyword arguments can be specified as. So here's another way of defining MLB team here:

>>>

>>> MLB_team = dict(

...   Colorado='Rockies',

...   Boston='Red Sox',

...   Minnesota='Twins',

...   Milwaukee='Brewers',

...   Seattle='Mariners'

...)

Once a dictionary has been established, you can show its contents, the same as you would do with a list. All three of the above definitions appear when displayed as follows:

```
>>>

>>> type(MLB_team)
<class 'dict'>
>>> MLB_team
{'Colorado': 'Rockies', 'Boston': 'Red Sox', 'Minnesota': 'Twins',
'Milwaukee': 'Brewers', 'Seattle': 'Mariners'}
```

Dictionary display entries in the order in which they were defined. But when it comes to retrieving them this is irrelevant. Numerical index is not used to access dictionary elements:

>>>

>>> MLB_team[1]

Traceback (most recent call last):

  File "<pyshell#13>", line 1, in <module>

    MLB_team[1]

KeyError: 1

The Dictionary Standards

Dictionary elements, of course, need to be somehow available. If you don't catalog them, then how do you get them?

For a dictionary, a value is retrieved by defining the corresponding key in square brackets([]):

>>>

>>> MLB_team['Minnesota']

'Twins'

>>> MLB_team['Colorado']

'Rockies'

If you are referring to a key not in the dictionary, Python makes an exception to:

>>>

>>> MLB_team['Toronto']

Traceback (most recent call last):

  File "<pyshell#19>", line 1, in <module>

    MLB_team['Toronto']

KeyError: 'Toronto'

To add an entry to an existing dictionary is simply to assign a new key and value to:

>>>

>>> MLB_team['Kansas City'] = 'Royals'

>>> MLB_team

{'Colorado': 'Rockies', 'Boston': 'Red Sox', 'Minnesota': 'Twins',

'Milwaukee': 'Brewers', 'Seattle': 'Mariners', 'Kansas City': 'Royals'}

If you want an entry updated, simply assign a new value to an existing key:

>>>

>>> MLB_team['Seattle'] = 'Seahawks'

>>> MLB_team

{'Colorado': 'Rockies', 'Boston': 'Red Sox', 'Minnesota':
'Twins',

'Milwaukee': 'Brewers', 'Seattle': 'Seahawks', 'Kansas
City': 'Royals'}

Use del statement to delete an entry, specifying the key to
delete:

```
>>>
```

>>> del MLB_team['Seattle']

>>> MLB_team

**{'Colorado':   'Rockies',   'Boston':   'Red   Sox',
'Minnesota': 'Twins',**

'Milwaukee': 'Brewers', 'Kansas City': 'Royals'}

Begone, Seahawks! Thou art an NFL team.

List Indices vs. Dictionary Keys

You may have found that when a dictionary is accessed
with either an unknown key or a numeric index, the
interpreter raises the same problem, KeyError:

```
>>>
```

```
>>> MLB_team['Toronto']
```

Traceback (most recent call last):

 File "<pyshell#8>", line 1, in <module>

    MLB_team['Toronto']

KeyError: 'Toronto'

>>> MLB_team[1]

Traceback (most recent call last):

  File "<pyshell#9>", line 1, in <module>

    MLB_team[1]

KeyError: 1

Actually it's the same mistake. In the latter case, [1] appears to be a numerical index, but not so.

In this guide you will see later that an object of any immutable type can be used as a key to the dictionary. There's therefore no reason you can't use integers:

>>>

>>> d = {0: 'a', 1: 'b', 2: 'c', 3: 'd'}

>>> d

{0: 'a', 1: 'b', 2: 'c', 3: 'd'}

>>> d[0]

'a'

>>> d[2]

'c'

The numbers in square brackets appear in the expressions MLB team[1], d[0], and d[2] as if they could be indices. But the order of the items in the dictionary has nothing to do with them. Python interprets them as keys to a dictionary. If you define the same dictionary in reverse order, using the same keys, you still get the same values:

```
>>>
>>> d = {3: 'd', 2: 'c', 1: 'b', 0: 'a'}
>>> d
{3: 'd', 2: 'c', 1: 'b', 0: 'a'}
>>> d[0]
'a'
>>> d[2]
'c'
```

The syntax may look similar but a dictionary cannot be treated like a list:

```
>>>
>>> type(d)
<class 'dict'>
>>> d[-1]
Traceback (most recent call last):
```

  File "<pyshell#30>", line 1, in <module>

    d[-1]

KeyError: -1

>>> d[0:2]

Traceback (most recent call last):

  File "<pyshell#31>", line 1, in <module>

    d[0:2]

TypeError: unhashable type: 'slice'

>>> d.append('e')

Traceback (most recent call last):

  File "<pyshell#32>", line 1, in <module>

    d.append('e')

AttributeError: 'dict' object has no attribute 'append'

Note: While access to items in a dictionary does not depend on order, Python guarantees the preservation of the order of items in a dictionary. When displayed, the items will appear in the order they've been defined, and iteration will also occur in that order through the keys. In the end, add items added to a dictionary. If items are deleted, it retains the order of the remaining items.

Very recently, you can only count on the preservation of order. It was added in version 3.7 as part of Python language specification. However, this was also true of version 3.6—by occurrence as a result of the implementation but not guaranteed by the language specification.

Building an Incremental Dictionary

Defining a dictionary using curly braces and, as shown above, a list of key-value pairs is fine if you know all the values and keys in advance. But what if you want to construct a dictionary on the fly?

You can begin by creating an empty dictionary, with empty curly braces defined. You can then add new keys and values, one by one:

```
>>>

>>> person = {}
>>> type(person)
<class 'dict'>
```

```
>>> person['fname'] = 'Joe'
>>> person['lname'] = 'Fonebone'
>>> person['age'] = 51
>>> person['spouse'] = 'Edna'
>>> person['children'] = ['Ralph', 'Betty', 'Joey']
>>> person['pets'] = {'dog': 'Fido', 'cat': 'Sox'}
```

Once this creates the dictionary, its values are accessed in the same manner as any other dictionary:

```
>>>
>>> person
{'fname': 'Joe', 'lname': 'Fonebone', 'age': 51, 'spouse':
'Edna',
'children': ['Ralph', 'Betty', 'Joey'], 'pets': {'dog': 'Fido',
'cat': 'Sox'}}


>>> person['fname']
'Joe'
>>> person['age']
51
```

```
>>> person['children']

['Ralph', 'Betty', 'Joey']
```

A supplementary index or key is needed to recover the values in the sublist or subdictionary:

```
>>>

>>> person['children'][-1]

'Joey'

>>> person['pets']['cat']

'Sox'
```

This example shows another dictionary feature: the values in the dictionary need not be of the same type. Some of the values are strings in person, one is an integer, one is a list, and another is a dictionary.

Just as the values in a dictionary need not be of the same type, so do not the keys:

```
>>>

>>> foo = {42: 'aaa', 2.78: 'bbb', True: 'ccc'}

>>> foo

{42: 'aaa', 2.78: 'bbb', True: 'ccc'}
```

```
>>> foo[42]

'aaa'

>>> foo[2.78]

'bbb'

>>> foo[True]

'ccc'
```

One of the keys here is an integer; one is afloat, and the other a Boolean. How that would be useful is not obvious, but you never know.

Notice how versatile dictionaries are to Python. In the MLB team, the same piece of information is kept for each of the different geographical locations (the name of the baseball team). Person, on the other hand, stores varying data types for one individual.

You can use dictionaries for a wide range of purposes because the keys and values that are allowed are so few limitations. But there are a few. Read on for more!

Limits on Dictionary Key

In Python it is possible to use almost any form of value as a dictionary key. You just saw this example, where objects like integer, float and boolean are used as keys:

```
>>>
```

```
>>> foo = {42: 'aaa', 2.78: 'bbb', True: 'ccc'}

>>> foo

{42: 'aaa', 2.78: 'bbb', True: 'ccc'}
```

You can also use included objects such as forms and functions:

```
>>>

>>> d = {int: 1, float: 2, bool: 3}

>>> d

{<class 'int'>: 1, <class 'float'>: 2, <class 'bool'>: 3}

>>> d[float]

2

>>> d = {bin: 1, hex: 2, oct: 3}

>>> d[oct]

3
```

There are however a few restrictions that dictionary keys must adhere to.

Second, a given key can only show up once in a dictionary. Do not require duplicate keys. A dictionary maps each key to a corresponding value, so mapping a particular key more than once doesn't make sense.

You noticed above that it does not apply the key a second time when you assign a value to an existing dictionary key, but then substitutes the original value:

```
>>>
>>> MLB_team = {
...     'Colorado' : 'Rockies',
...     'Boston'   : 'Red Sox',
...     'Minnesota': 'Twins',
...     'Milwaukee': 'Brewers',
...     'Seattle'  : 'Mariners'
... }

>>> MLB_team['Minnesota'] = 'Timberwolves'
>>> MLB_team
{'Colorado': 'Rockies', 'Boston': 'Red Sox', 'Minnesota':
'Timberwolves',
'Milwaukee': 'Brewers', 'Seattle': 'Mariners'}
```

Likewise, if you specify a key a second time during initial dictionary formation, the second occurrence will override the first one:

```
>>>

>>> MLB_team = {

...     'Colorado' : 'Rockies',

...     'Seattle'  : 'Mariners',

...     'Minnesota': 'Twins'

...     'Boston'   : 'Red Sox',

...     'Minnesota': 'Timberwolves',

...     'Milwaukee': 'Brewers',

... }

>>> MLB_team

{'Colorado': 'Rockies', 'Boston': 'Red Sox', 'Minnesota':
'Twins',

'Milwaukee': 'Brewers', 'Seattle': 'Mariners'}
```

Begone, timberwolf! You 're an NBA Team. Sort of.

Secondly, a key to a dictionary must be of an immutable kind. You've already seen examples of where several of the immutable types you know — integer, float, string, and boolean — have served as dictionary keys.

Even a tuple can be a dictionary element, since tuples are immutable:

```
>>>
>>> d = {(1, 1): 'a', (1, 2): 'b', (2, 1): 'c', (2, 2): 'd'}
>>> d[(1,1)]
'a'
>>> d[(2,1)]
'c'
```

(Remember from the discussion on tuples that one reason for using a tuple rather than a list is that there are circumstances under which an immutable form is needed.

Neither a list nor any other dictionary, however, may serve as a dictionary base, as lists and dictionaries are mutable:

```
>>>
>>> d = {[1, 1]: 'a', [1, 2]: 'b', [2, 1]: 'c', [2, 2]: 'd'}
Traceback (most recent call last):
  File "<pyshell#20>", line 1, in <module>
    d = {[1, 1]: 'a', [1, 2]: 'b', [2, 1]: 'c', [2, 2]: 'd'}
TypeError: unhashable type: 'list'
```

Technical Note: Why does the error message say "unhashable"?

Technically, saying an object must be immutable to be used as a dictionary key is not correct. More precisely, an object must be hashable, which means a hash function can be passed on to it. A hash function takes arbitrary size data and maps it to a relatively simpler fixed-size value called a hash (or simply hash) value that is used to look up and compare the table.

The built-in hash) (function of Python returns the hash value for an object that can be hashed, and raises an exception for an object that is not:

```
>>>

>>> hash('foo')

11132615637596761

>>> hash([1, 2, 3])

Traceback (most recent call last):
  File "<stdin>", line 1, in <module>

TypeError: unhashable type: 'list'
```

All the built-in immutable types you've heard about up to now are hashable, and there aren't the mutable container styles (lists and dictionaries). You should think of hashable and immutable as more or less interchangeable for present-day purposes.

You'll find mutable objects that are also hashable in future guides.

Limits on Dictionary values

Conversely, dictionary values are not restricted. Literally, absolutely none. A dictionary value can be any type of object that is supported by Python, including mutable types such as lists and dictionaries, and user-defined objects that you'll learn about in future guides.

```
>>>
>>> d = {0: 'a', 1: 'a', 2: 'a', 3: 'a'}
>>> d
{0: 'a', 1: 'a', 2: 'a', 3: 'a'}
>>> d[0] == d[1] == d[2]
True
```

Operators and Built-in Functions

Many of the operators and built-in functions that could be used for strings, lists, and tuples have already become known to you. Some of those also work for dictionaries.

The in and not in operators, for example, return True or False according to whether the stated operand appears as a key in the dictionary:

```
>>>
>>> MLB_team = {
...     'Colorado' : 'Rockies',
...     'Boston'   : 'Red Sox',
...     'Minnesota': 'Twins',
...     'Milwaukee': 'Brewers',
...     'Seattle'  : 'Mariners'
... }

>>> 'Milwaukee' in MLB_team
True
>>> 'Toronto' in MLB_team
False
>>> 'Toronto' not in MLB_team
True
```

Together with the short-circuit test, you can use the in operator to avoid making an error while attempting to reach a word which is not in the dictionary:

```
>>>
>>> MLB_team['Toronto']
Traceback (most recent call last):
  File "<pyshell#2>", line 1, in <module>
    MLB_team['Toronto']
KeyError: 'Toronto'


>>> 'Toronto' in MLB_team and MLB_team['Toronto']
False
```

In the second case, the term MLB team['Toronto] 'is not evaluated due to short-circuit evaluation, so the KeyError exception does not occur.

The function len () returns a dictionary with the number of key-value pairs:

```
>>>
>>> MLB_team = {
...     'Colorado' : 'Rockies',
```

...   'Boston' : 'Red Sox',

...   'Minnesota': 'Twins',

...   'Milwaukee': 'Brewers',

...   'Seattle' : 'Mariners'

... }

>>> len(MLB_team)

5

Methods built-in Dictionary

As with strings and lists, several built-in methods can be invoked on dictionaries. The list and dictionary methods share the same name in some cases. (In the discussion about object-oriented programming, you'll see that having methods with the same name is perfectly acceptable for different types.)

An overview of methods which apply to dictionaries is as follows:

D.clear)-Clear)

Clears a dictionary.

D.clear()empties dictionary d of all pairs with the key value:

>>>

```
>>> d = {'a': 10, 'b': 20, 'c': 30}
>>> d
{'a': 10, 'b': 20, 'c': 30}


>>> d.clear()
>>> d
{}
```

d.get(<key>[, <default>])

Returns the value for a key If it exists in the dictionary.

The Python dictionary.get) (method provides a convenient way to get a key value from a dictionary without pre-checking if the key exists, and without raising an error.

D.get(<key >) checks for < key > in the dictionary d and returns the associated value if found. If < key > is not found then None returns:

```
>>>
>>> d = {'a': 10, 'b': 20, 'c': 30}
```

```
>>> print(d.get('b'))

20

>>> print(d.get('z'))

None
```

If < key > is not found, and the optional argument < default > is specified, this value is returned instead of None:

```
>>>

>>> print(d.get('z', -1))

-1
```

d.items()

Returns a list of key-value pairs in a dictionary.

D. items () returns a list of the key-value pair tuples in d. In each tuple the first item is the key, and the second item is the value of the key:

```
>>>

>>> d = {'a': 10, 'b': 20, 'c': 30}

>>> d

{'a': 10, 'b': 20, 'c': 30}
```

```
>>> list(d.items())
[('a', 10), ('b', 20), ('c', 30)]
>>> list(d.items())[1][0]
'b'
>>> list(d.items())[1][1]
20
```

D. keys ( )

Returns a list of keys in a dictionary,

D.keys () returns all keys listed in d:

```
>>>
>>> d = {'a': 10, 'b': 20, 'c': 30}
>>> d
{'a': 10, 'b': 20, 'c': 30}


>>> list(d.keys())
['a', 'b', 'c']
```

D. values ( )

Returns a list of values in a dictionary.

D.values() returns a list of all values in d:

```
>>>
>>> d = {'a': 10, 'b': 20, 'c': 30}
>>> d
{'a': 10, 'b': 20, 'c': 30}
```

```
>>> list(d.values())
[10, 20, 30]
```

Any duplicate values in d will be returned as many times as they occur:

```
>>>
>>> d = {'a': 10, 'b': 10, 'c': 10}
>>> d
{'a': 10, 'b': 10, 'c': 10}
```

```
>>> list(d.values())
[10, 10, 10]
```

Technical Note: The methods. 'items(), 'keys), (and. 'values) (actually return something called an object of view. An object with a dictionary view is more or less like a window on the keys and values. For practical purposes, these methods can be thought of as returning lists of the keys and values of the dictionary.

D.pop (<key >, [< default >])

Removes a key from a dictionary, and returns its meaning if it is present.

If < key > is in d, d.pop(<key >) will delete < key > and return its associated value:

```
>>>
>>> d = {'a': 10, 'b': 20, 'c': 30}


>>> d.pop('b')
20
>>> d
{'a': 10, 'c': 30}
```

d.pop(<key>) raises a KeyError exception if <key> is not in d:

```
>>>
>>> d = {'a': 10, 'b': 20, 'c': 30}
```

```
>>> d.pop('z')

Traceback (most recent call last):

  File "<pyshell#4>", line 1, in <module>

    d.pop('z')

KeyError: 'z'
```

If < key > is not in d and the optional argument < default > is specified, the value will be returned and no exception will be raised:

```
>>>

>>> d = {'a': 10, 'b': 20, 'c': 30}

>>> d.pop('z', -1)

-1

>>> d

{'a': 10, 'b': 20, 'c': 30}
```

D.popitem ()

Removes from a dictionary a key-value paire.

D.popitem() extracts and returns the last key-value pair added from d as a tuple:

```
>>>

>>> d = {'a': 10, 'b': 20, 'c': 30}
```

```
>>> d.popitem()
('c', 30)
>>> d
{'a': 10, 'b': 20}


>>> d.popitem()
('b', 20)
>>> d
{'a': 10}
```

If d is empty d.popitem() will create an exception to KeyError:

```
>>>
>>> d = {}
>>> d.popitem()
Traceback (most recent call last):
  File "<pyshell#11>", line 1, in <module>
    d.popitem()
KeyError: 'popitem(): dictionary is empty'
```

Note: popitem() will return an arbitrary (random) key-value pair in Python versions less than 3.6, since Python dictionaries were unordered prior to version 3.6.

The.update(<obj >)

Combines a dictionary with another dictionary, or a key-value pair iterable.

If < obj > is a dictionary, d.update(<obj >) merges the < obj > entries into d. For every single key in < obj >:

• If the key does not appear in d, the key-value pair < obj > is added to d.

• If the key already exists in d, the corresponding value in d for that key is updated to < obj >.

Here is an example which shows two merged dictionaries:

```
>>>

>>> d1 = {'a': 10, 'b': 20, 'c': 30}

>>> d2 = {'b': 200, 'd': 400}


>>> d1.update(d2)
>>> d1

{'a': 10, 'b': 200, 'c': 30, 'd': 400}
```

In this example , key 'b' already exists in d1, thus updating its value to 200, the d2 value for that key. There is no key 'd 'in d1, however, so that key-value pair is added from d2.

<obj> can also be a sequence of key-value pairs, similar to those used to define a dictionary by the dict). For example, a list of tuples can be specified as <obj>:

```
>>>
>>> d1 = {'a': 10, 'b': 20, 'c': 30}
>>> d1.update([('b', 200), ('d', 400)])
>>> d1
{'a': 10, 'b': 200, 'c': 30, 'd': 400}
```

Or the values to merge can be specified as a list of keyword arguments:

```
>>>
>>> d1 = {'a': 10, 'b': 20, 'c': 30}
>>> d1.update(b=200, d=400)
>>> d1
{'a': 10, 'b': 200, 'c': 30, 'd': 400}
```

Lists and dictionaries are two of the most widely used styles of Python. They have several similarities, as you have seen, but differentiate in how their elements are accessed. List elements are accessed by order-based numeric index, and primary links to dictionary elements

Because of this difference, lists and dictionaries tend to suit various circumstances. You should have a good feeling now that would be best for a given situation if either.

You'll read more about the Python sets next. The collection is another type of composite data, which is somewhat different from a list or dictionary.

CHAPTER 5

COMPUTER AND PYTHON PROGRAMMING

Computer programming is the process that experts use to write code that teaches how to perform a computer, application, or software program. Computer programming, at its most simple, is a series of instructions for executing specific behavior. Computer programmers generate instructions for a computer to process by writing and testing code to allow the successful operation of applications and software programs.

Computers can do incredible things, from basic laptops capable of simple word processing and spreadsheet functions to incredibly complex super-computers completing thousands of financial transactions a day and controlling the infrastructure that makes modern life possible. But no machine can do anything unless it is instructed by a computer programmer to act in some way. So it's all about computer programming.

Computer programming, at its most basic, is little more than a set of instructions for facilitating specific actions. Computer programming can be as easy as adding two numbers, depending on the specifications or purposes of such instructions. It could also be as complex as reading temperature sensor data to adjust a thermostat, sorting data to complete critical reports or intricate scheduling, or taking players through multilayered worlds and game challenges.

Senior Dean of STEM programs at Southern New Hampshire University (SNHU), Cheryl Frederick said computer programming is a collaborative activity, with several programmers contributing to the creation of a piece of software. Some of that evolution may last for decades. For example, programmers have been tweaking and improving for years for software like Microsoft Word released in 1983.

"The expectation is that the computer program will become such a widely accepted system that it requires long-term support, particularly for expanding its current functionality," said Frederick. "The terms computer programming and computer software are interchangeably used, except that software can be used instead.

What does a programmer do for a computer?

Computer programmers generate instructions for a computer to follow writing and testing code to allow applications and software programs to run effectively. Computer programmers utilize specialized languages to communicate with machines, applications, and other systems to get computers and computer networks to perform a series of specific tasks.

Many programming languages exist, but some have emerged as the most popular. Based on a review of 100,000 programmers, industry blog The Crazy Programmer recently listed the top 10 programming languages to be used in 2018. They cover:

- JavaScript

- C

- TypeScript

- SQL

- Java

- C++

- Python

- C#

- PHP

- Ruby

O*Net compiled some of the common tasks that a computer programmer has to master online and include:

- Writing programming code on your computer.

- Working with others to resolve IT problems.

- Solving problems with computer software.

- Modifying performance-enhancing software programs.

- To test the performance of the program.

How to Become a Programmer

Many programmers start as self-educated enthusiasts. Dr. Ed Lavieri began as an autodidactic gamer and worked in the Navy for 25 years before becoming a full-time instructor. As the technical program facilitator for game programming and development, he said, "When you have the core knowledge, computer programming becomes fun."

"But, you can not rest on the information you have learned," said Lavieri. "A degree is a guide to past knowledge.

Certifications, skills, degrees, an e-portfolio – all of these will help you get your foot in the door; however, if you want to be the one to create Windows 11, you have to get a wide range of experience and take advantage of opportunities as they come.

Frederick consented, before turning to education, she held jobs with the Defense Department and in the financial and telecommunications sectors. "It takes a lot of grit, and earning a degree isn't enough; you need workplace experience," she said. "We 're giving students a foundation – based on data structures, algorithms, math, and logical engineering – but you need to be able to plan, write, design, test, and manage software. You will learn at least two or three programming languages, C++ skills, and JAVA.'

However, beyond the classroom and experiential learning, computer programmers must understand that the first time a program is written, it never works. "This field demands patience and the ability to troubleshoot and make mistakes. You need to be an apprenticeship worker, have the motivation to learn on your own, be self-disciplined, be able to brainstorm with others, and have a lot of hands-on practice, "said Frederick. "You have to be a practitioner and adapt to the trend."

"Game development – a multi-billion dollar business that needs much more than graphics and sound – is one of the toughest programs there are," Lavieri said. "But every industry needs programmers without exception, from health care and real estate to banking, travel, and any other sector."

While working towards completing a degree in computing science, students are encouraged to create a portfolio of their software work. "Although this portfolio is not graded, students can share it as proof of coding capabilities with potential employers," Frederick said.

"The entire degree program provides students with extensive exposure and expertise in traditional and trending technologies, including specialties such as computer graphics, software testing and writing code for commonly used programs, as well as deeper, more specific skills."

Certifications for Computer Programming To a Degree

According to the USA, most computer programming positions require at least a bachelor's degree. There is also a Bureau of Labor Statistics and other specialized degree programs. There are hundreds of technical and non-profit certifications available, in addition to those academic avenues. BLS notes that there are certifications for specific programming languages and that some employers may require programmers to be certified in the products used by the company.

"The entire degree program provides students with extensive exposure and expertise in traditional and trending technologies, including specialties such as computer graphics, software testing and writing code for commonly used programs, as well as deeper, more specific skills."

Some of the professional certifications available include:

Professional Associations – Comptia's Security+, Software Development Associate Certification, Comptia's A+ Certification, Comptia's Linux+

Microsoft – Certified Solutions Associate Windows Server, Certified Solution Developer for Web Applications,

Nonprofit – Certified Information Security Manager, Certified Information Systems Security Professional, Certified Secure Software Lifecycle Professional Credential

CISCO – Certified Network Associate, Certified Network Profession Routing and Switching, Certified Network Associate Security Credential

STARTING WITH PYTHON

If you've already decided to start learning Python but don't know where to start, then this is your spot. It can be hard to get started. Should you download Python 2.7 or just download Python 3.6, let's say? So we've compiled extensive information and practical tips in this section, which will help you find your way. You'll be given an overview of the differences and guide you through the installation process between Python 2 and Python 3. Then, we will review the best libraries and IDEs for Python and their use cases. What's more, here you will find actionable steps to begin learning Python, as well as a list of good resources you can use. Lastly, we'll show you how to get support from seasoned Python users if you need it. But first things:

Which version of Python do you want to learn?

If you've tried downloading Python, the two seemingly equally important versions-2.7 and 3.7 (as of writing this article) may have surprised you. So now, you probably wonder: Should I learn Python 2, or should I learn Python 3?

This version was developed primarily for fixing problems that existed in Python 2. However, the nature of these changes is such that Python 3 proved incompatible with Python 2! As a result, migrating their project to 3.x required lots of adjustments and effort for any organization that used the Python 2.x version. Many companies, therefore, decided to continue to use version 2 and to develop further libraries for it. And so, it continued support for Python 2.

But, eventually, all things have to end.

Python 2 retired January 1, 2020. There will be no support and updates for this version after this date, and anybody using it will need to switch to Python 3 ASAP. That's not a simulation-the clock is ticking literally.

So Python 3 is the simple path for the future and the version that should be chosen by any beginner. If your company has projects written in Python 2 or, perhaps, you need to work with a third-party library that only supports Python 2, you may need to learn the differences between the two versions. Here are some (there's a lot more to get you started from where this came from):

Feature to compare	Python 2	Python 3
The print statement	print "Hello, world"	Print ("Hello, world")
Division of integers	In this version, 5/2 would perform an integer division with a result of 2. If you need to get the actual answer, you need to write 5.0/2.0	Here, 5/2 results in 2.5, as expected. This makes your code clearer and understandable, with fewer bugs.

Feature to compare	Python 2	Python 3
Unicode	In Python 2, strings are stored as ASCII by default. You need to add "u" to convert them to Unicode	Text strings are Unicode by default.

PYTHONS IDES FOR USE IN RESEARCH AND DATA PROCCESSING SPYDER

Spyder is an open-source, lightweight IDE designed and built specifically for use in the scientific and data analysis. If you've never been working with an IDE, it might be a perfect first stop because the curve for learning is smooth.

Spyder is included in the Anaconda distribution, alongside the data science and machine learning libraries – NumPy, SciPy, sci-kit-learn, Pandas, Matplotlib, and so on. And, in terms of deployment, the optimal way to do it is by Anaconda.

Spyder provides most of the popular IDE features you would expect, such as a powerful syntax-highlighting code editor, code completion, and even integrated plugin documentation.

It also features a few extremely helpful features-a graphical variable explorer and IPython (interactive Python) console.

The variable explorer contains all the data (variables) from your program and displays it right inside your IDE using a table-based layout. This enables you to interact and modify it in just a few clicks on the fly, plot the histograms and time series, sort collections, and more.

Another interesting thing you might find is that the IPython console greatly benefits beginners and data scientists by allowing them to execute a single line of code and visualize the data. And that is great both for educational and debugging purposes.

Essentially, Spyder is good for beginners as well as professionals, but we realize that some veterans may feel it is too basic and lack some more advanced features.

Jupyter Notebook

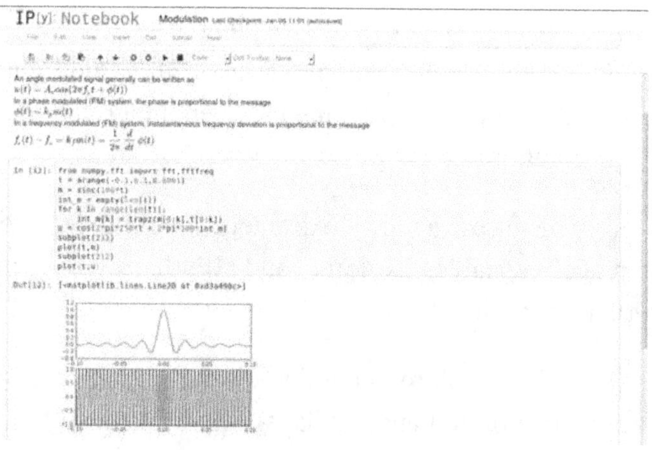

The Jupyter Notebook appears to support markdowns, allowing you to add text and other HTML components between your code lines, such as images and videos. Thanks to Jupyter, your code can be easily viewed and edited to create compelling presentations. For example, you can use data visualization libraries such as Seaborn and Matplotlib, and show your graphs where your code is in the same document. You can also export your final work to PDF and HTML files, or simply export it as a .py file.

In 2014, Jupyter Notebook was born from IPython. It is a web application based on the structure of the server-client and allows you to create and manipulate notebook documents-or just notebooks.

Jupyter offers a user-friendly, immersive data science platform across several programming languages that acts not only as an IDE but as a presentation or educational resource as well. It is perfect for those who just start data science!

COMMON USE OF PYTHON IDES

PyCharm

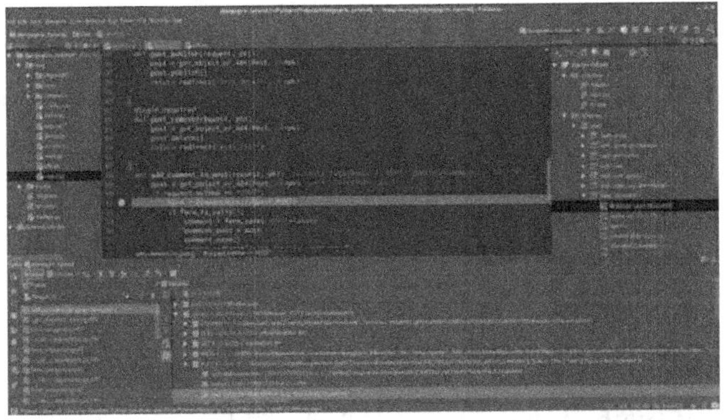

PyCharm is a complete Web development IDE, suitable for small and large projects. It's an IDE made by the folks at JetBrains, a company known for creating fantastic software development tools for professional developers. If you have experience using another of JetBrains' IDEs, it is perfect for you because the interface and features are similar.

Also, if you like the distribution of IPython or Anaconda, it's good for you to know that PyCharm combines their tools and libraries like NumPy and Matplotlib. And that helps you to collaborate with viewers from the variety and immersive plots.

Well, there are two versions of PyCharm to consider in terms of installation:

• Community – free, lightweight, open-source version, good for Python and scientific development;

• Professional – paid version (53 EUR / year, 2 years later), full-featured IDE with Web development support as well.

PyCharm 's downside is that it can be quite heavy and resource-intensive. Therefore, computers with a small amount of RAM (usually less than 4 GB) may not be the best choice, because it will lag.

Other than that, PyCharm provides all the major features out of the box that a good IDE should deliver. Plus, it is fully customizable and has plenty of plugins for additional features.

Visual Studio Code (VSCode)

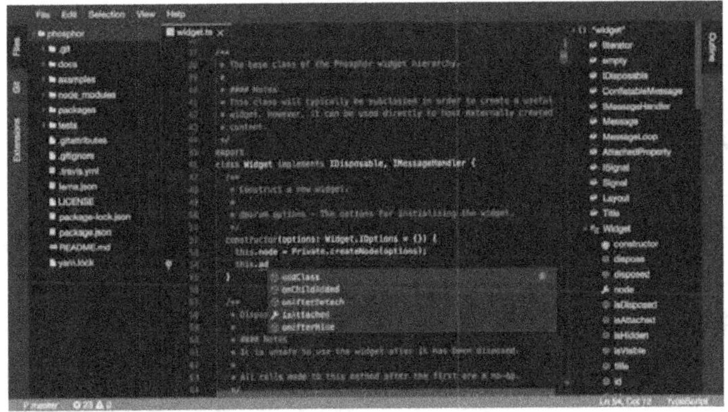

VSCode is a lightweight but powerful open-source editor that can be accessed on any platform (Mac, Windows, Linux). It has a rich built-in support ecosystem and programming language extensions. The editor receives updated monthly with new bug fixes and features. This, and the active community that generates useful plugins for different usage cases, make VSCode a favorite of programmers when it comes to developing Python.

What do you get from the extension?

- Auto-completion with IntelliSense;

- Linting;

- Refactoring commands.

- Unit testing;

- Code formatting;

- Script debugging;

- Easy switch and automatic activation between Python environments;

VSCode extensions include other handy features as well as programming language capabilities, such as keymaps, language packs, and UI themes. Extensions and themes are super easy to install and accessible. So, yes, you definitely should put this IDE on your radar once you get started working with Python.

Atom

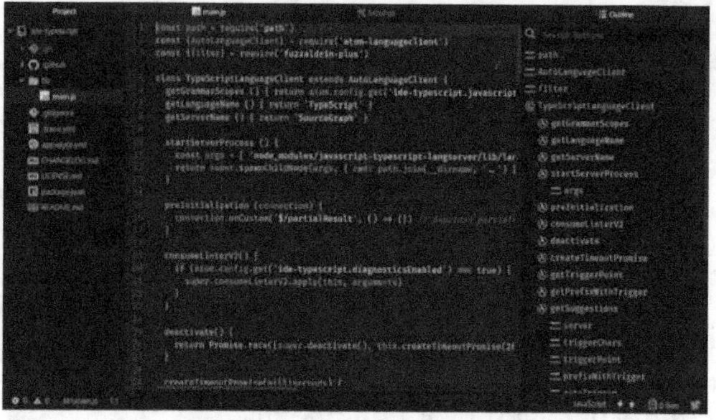

Atom is an open-source code editor (not a full IDE), developed by GitHub for use in Python development. It is highly customizable, allowing you to install packages according to your needs and to adjust the user interface whenever you want.

Atom can be easy to use, and learn easily. It gives GitHub great courtesy in supporting. Also, you can view your results in Atom itself without opening any other window or panel. Additionally, you also have a "Markdown Preview Plus" plugin. It provides you with built-in editing support, and by viewing Markdown files, you can open a preview, render LaTeX equations, and much more!

Honorable Mentions

• Eclipse + PyDev-Eclipse is a common IDE used primarily for Java code. It can be configured for Python though using the PyDev plugin.

It (Online Compiler)-You can use repl.it if you want to start writing Python code without investing time to install Python and set up a development environment. The website provides numerous languages with online IDE, including Python shell. To get going, you need only an internet connection and a browser.

• Thonny – Thonny is billed as an IDE for beginners, a recent addition to the Python IDE Family. Written and maintained by the University of Tartu, Estonia, Institute of Computer Science, it is specifically designed for teaching and learning programming (in particular, Python). Thonny comes with its bundled version of Python by default, so you don't have to install anything else. It has a user interface that makes it easy for beginners to work with. It also features a simple debugger that allows you to execute your program step-by-step and enables you to see how Python evaluates the written expression internally.

What are the top Libraries and Frameworks in Python?

The Python libraries contain different tools to help us program. We may find a difficult problem to solve, but other people may have encountered it before us and have written code to handle it. That code is probably awaiting us to use it somewhere, in some library. There are numerous different libraries for Python, but here we will list only those that we think you'll probably need to use.

Standard python library

Packed with Python itself, this is the library. It's very extensive, and it offers a wide range of facilities that you can check in the documentation. The library contains embedded modules (written in C). Those provide device functionality access, such as the I / O file (that would otherwise be inaccessible to Python programmers). It also opens the door to Python-written modules that provide standardized solutions to several common programming problems. Some of these modules are explicitly designed by abstracting platform-specifics into platform-neutral APIs to encourage and enhance the portability of Python programs.

Python machine learning libraries, scientific use, and data analysis

Python is usually used for data analysis and scientific simulations, as we have mentioned before. It owes its popularity to some extent to the vast array of libraries that you can select from. Here is what Oguzhan Gencoglu, co-founder and head of AI at Top Data Science, says: "We create mostly in Python, and it's important that scientific libraries like NumPy, Pandas, and Scipy. We do use Scikit-learn / Xgboost / GPy when it comes to conventional ML. Continuing the list is essential for our NLP projects: Gensim and Nltk. Of course, we prototype in Keras / TensorFlow for deep learning and deploy in TensorFlow, If it is a collaboration between industry. We also used PyTorch for collaborations with the research.

That said, below is a list of some of Python's most significant libraries:

- NumPy

This is an important library for numerical calculations and the basis for many other libraries that are more advanced. It provides extensive N-dimensional linear algebra functions and array interface, which are faster order of magnitude and more efficient memory than normal Python lists. For this reason, simulations and machine learning are essential.

NumPy is part of kit Anaconda.

- The Scientific Python (SciPy)

SciPy is a library of algorithms developed to perform complex mathematical analysis. It uses NumPy arrays as its basic data structure (thus it's fast) and can perform tasks such as integration, solving differential equation, signal processing, and optimization.

SciPy is made available through the Anaconda package.

• Matplotlib

Matplotlib is a library designed to create plots, graphs, and other visual data representations. For any data scientist or data analyzer, it is vital.

Matplotlib is a part of the Anaconda package as well.

• TensorCurrent

This is an open-source, machine learning library developed by the Google Brain Team and released in 2015 for the general public. It is used in almost every machine-learning program on Google. The number of TensorFlow applications is infinite, and that is its beauty. It is designed for speed, it is very versatile, and it can run both on CPU and GPU. It also comes with nice tools for visualization. TensorFlow is a piece of the Anaconda package, excluding automatic installation-a manual installation is required. This can be done by opening the prompt anaconda and type "conda install TensorFlow"

• Pandas

Pandas is a library for data processing, offering high-level data structures and a wide variety of resources for manipulating them. One of the great highlights of this library is its ability to use one or two commands to translate complex operations using data. Pandas have several built-in methods for grouping, combining, and filtering data, as well as functionality in the time series.

Pandas is part of package Anaconda.

• Scikit-learn

Scikit-learn is a Python library related to NumPy and SciPy. It is considered to be one of the best libraries for working with complex data, since it includes various algorithms for implementing standard machine learning and data mining tasks, such as reducing dimensionality, classification, regression, clustering, and modeling.

Scikit-learn is part of package Anaconda.

PyTorch

This is another open-source machine learning library that allows developers to perform GPU accelerated tensor computations, build dynamic computational graphs, and automatically calculate gradients. It was launched in 2017, and since its inception, the library is gaining popularity and attracting an increasing number of developers in machine learning. Despite being fairly young and not possessing TensorFlow 's outstanding visualization tools, PyTorch and TensorFlow. are in direct competition.

PyTorch is part of the Anaconda package; however, it requires manual installation – open anaconda prompt and enter "Conda PyTorch -c PyTorch installation."

• Keras

Keras provides a simpler mechanism for expressing neural networks, as well as some of the best tools for compiling models, processing datasets, graph visualization, and more. It focuses on being user-friendly, flexible, and expandable, but speed is sacrificed.

Keras is a favorite among researchers on deep learning, coming in at # 2. Researchers at major research organizations, in particular NASA and CERN, have also embraced it.

Keras is not a member of kit Anaconda.

• wxPython

WxPython is an open-source wrapper for WxWidgets (previously called WxWindows) cross-platform GUI library and is implemented as a Python extension module. With WxPython, you can create native Windows, Mac OS, and Unix applications as a developer.

Libraries with Python GUI

GUI libraries (Graphical User Interface) allow programmers to build glamorous applications. Python is an interactive language, so programming a GUI framework is not much of a difficult task to get started. In fact, in this sector Python has a diverse range of options:

• TKTER

This is commonly bundled with Python, using Tk, which is the main GUI framework for Python. Its simplicity and graphical user interface make it popular. It is open-source and under the Python License. One of the benefits of choosing Tkinter is the abundance of resources, both codes, and reference books, since it comes by default.

• PyQT

PyQT is one of the preferred cross-platform Python bindings that implement the Qt library for the Qt application development framework (owned by Nokia). Unix / Linux, Mac OS X, Sharp Zaurus, and Windows currently offer PyQT. It combines Python's Best with Qt. And it's left to the programmer to decide whether by coding or using Qt Designer to create visual dialogs to create a program. It is available in both GPL and commercial licenses. Although certain features in the free version might not be available, if the program is open source, then you can utilize it under the free license.

• PyGUI Inc.

PyGUI is a cross-platform graphical application framework designed for Unix, Macintosh, and Windows. PyGUI is by far the easiest and lightweight of them all compared to some other GUI implementations, as the API is strictly in line with Python. PyGUI adds much fewer code between the GUI interface and the Python program. Thus, the program display usually shows the platform's natural GUI.

PYTHON WEB DEVELOPMENT FRAMEWORK

The popularity of Python frameworks appears to only increase, given how dynamic web development has become. This language of programming, which is object-oriented, strongly written, interpreted, and interactive, is very simple to understand. Besides, with its easy-to-read syntax and simple compilation function, it effectively reduces development time.

Python offers developers a wide variety of software. Some popular Web frameworks are listed below:

Django

Django is a free, open-source framework allowing developers to effectively and quickly develop complex code and applications. This high-level system streamlines the creation of web applications by giving various features. It has an assortment of libraries and underlines efficiency, fewer coding needs, and component reusability.

CherryPy

CherryPy is an open-source Web development framework for Python which implements its multi-strung server. It can continue to run on any Python supporting work framework. CherryPy features include a thread-pooled web server, setup system, and frame module. A moderate Web platform allows you to use any form of data access, templating, etc. technology. However, it can do anything a web server can, for example, manage sessions, static uploads, file uploads, cookies, etc.

Pyramid

Pyramid is a framework which underpins validation and guidance. It's amazing to grow huge web applications, like CMSs, and it's valuable for prototyping an idea and chipping away developers at API projects. Pyramid is adaptable and can be utilized for both simple and challenging projects. Pyramid is enriched with functionality that does not drive a particular way to complete items, lightweight without leaving any of you as the device grows. Because of its transparency and measured quality, it is the most valued Web framework among experienced Python developers. A moderate team and tech giants such as Mozilla, Yelp, SurveyMonkey, and Dropbox have used it.

Flask

Flask is an open Python application under the BSD license, which is inspired by the Sinatra Ruby project. Its principal purpose is to help develop a strong base of web applications. Developers can create backend frameworks whichever way they need, but it was designed for open-ended applications. Big companies, including LinkedIn and Pinterest, used Flask. Flask is better suited for small and simple tasks, compared to Django. So you can expect the development of a web server, support for the Google App Engine as well as in-built unit testing.

Bottle

Another interesting Web framework for Python is Bottle, which falls under the small-scale framework's class. It was originally developed for the creation of Web APIs. Bottle also tries to execute everything in a single document, which should give you a brief overview of how small it is intended to be. The functionalities of the out-of-the-box include templating, services, guiding, and some fundamental abstraction over the WSGI norm. Like Flask, you'll be coding a lot closer to the metal than a full-stack framework. Netflix has been using Bottle to create web interfaces, no matter what.

Where to start?

Now you probably know which libraries you 'd need for future projects. That's all good, but in Python, you still have to learn how to code. Also, presumably, in this section, you expect us to give you information about which tutorials to start, or which textbooks to choose from.

We inform you that this is your first (of many) disappointment on the path to mastering Python (or any other programming language). It's not that there are no good free guides – there are plenty out there, and some are great.

Unfortunately, however fantastic a tutorial can be, something would likely have been left out, or you would have made a mistake that is not explained therein. Then what? The solution is to check in with the most vital tool available to you-the internet. The thing is, most of the learning process would likely happen over Google in programming. For a programmer, it is a must to be able to find important information online. So, that's why we encourage you now to find your Python resources. Hmm ... do I hear a glimmer of despair?

We're not going to leave you totally to your own devices, right. Here are a few actionable steps you can take to get Python started.

What are the Applications of Python?

Python has grown in popularity and is now widely used in all kinds of applications. It is often referred to as "the easiest language to learn in programming." Python is used to develop video players such as YouTube, power applications such as Instagram, test microchips at Intel, run a Google search engine, and even power transactions at the New York Stock Exchange (NYSE). And you know that when it maintains a stock exchange system, a programming language is very powerful. Indeed, when they program their equipment and space machinery, NASA uses Python too. Well, is it something not? Let's look further into some of those applications.

PYTHON FOR DATA SCIENCE

Python has undergone a recent increase in popularity across various industries, primarily due to its libraries of data science.

Python offers a wide variety of data processing applications in both the business and academia. For a large part of the data science community, it is the language of choice. Why? For what? Thanks to the various resources that make it simple to work with enormous datasets and gain important insights.

Python libraries, such as Pandas and NumPy, are widely used to collect, process, and clean data sets, as well as to use mathematical algorithms for the benefit of users. The robust visualizations that come with Seaborn and

Matplotlib are another aspect that makes Python a great tool for data scientists.

PYTHON FOR MACHINE LEARNING

If you haven't already learned of Machine Learning (ML), here is a brief introduction. ML refers to a computer 's ability to "learn" from training data (fitting a model to the input) to make predictions (such as how likely it is for a customer to purchase this product based on past purchase data). It's widely used by many companies and in "smart" algorithms (where it uses large datasets).

There are indeed plenty of good reasons for companies to use Python.

Its libraries, such as TensorFlow, Scikit-learn, and NLTK, are widely used to predict trends such as customer satisfaction, stock projections, etc. TensorFlow has a very interesting story when talking about libraries. It was developed for in-house use by Google, but published as an open-source library in 2015. But in section 2.4, we'll get back to it where we'll talk more about the Python libraries.

This side of Python will improve even more, given the increasing importance of machine learning and artificial intelligence nowadays.

PYTHON FOR WEB DEVELOPMENT

HTTP programming (or, as we call it more commonly, web development) is no longer simply done with JavaScript. Python also plays a major part in it. So we'll introduce you to a major Python framework, called Django, in this section. Additionally, we'll see some of the big companies that use Django for their web services.

Django is a comprehensive Web framework. Now, what exactly does this mean? Full-stack frameworks give you everything you need to construct a complete web application. That includes web service, database management, and even generation of HTML. So, it's not a surprise that Django is the foundation of sites and services such as The Guardian, the New York Times, Instagram, and Pinterest. It looks like Django is the way to go when you're building large and complex web applications.

But what if you want simple Web applications to be created?

Let's use Python-Flask to sneak into another web content creation option. Flask is a web framework that is micro- and lightweight. That means it's more intuitive and user-friendlier. If you want to perform more complicated tasks, you may need a third party framework. But remember this: Flask overtook Django by 2 percent in 2018, with 47 percent of users opting for Flask and 45 percent opting for Django. This shows that the web development industry has shifted to smaller systems, microservices, and networks that are "serverless." It might be the perfect time to consider joining on the effort.

PYTHON FOR GAME DEVELOPMENT

We 're not going to sugarcoat it, that's where there is a lack of Python in the competition. Though it has game development libraries such as PyGame and PyOpenGL, they often prove to be insufficient to provide sufficient functionality. While generally good at creating 2D games, game developers consider it slow compared with compiled languages like C # and C++. Moreover, it's probably not the best graphics-wise choice, and it doesn't offer rich development tools or an editor. This, combined with the fact this Python is a high-level, interpreted language, makes game developers a not-so-attractive choice.

However, thanks to the fast speed of writing code in it, Python is extremely helpful when prototyping. Essentially, it is useful to build a prototype when deciding whether to invest time and resources in developing a game-a proof of concept. And that's where Python comes in, as it can produce a simple 2D prototype that works much faster than other languages.

Python also plays a major role in the online gaming culture. Notable examples of games created with its support are Eve Online (uses Stackless Python), Civilization IV (uses Python for logic and server controls), and World of Tanks (uses Python for some of its internal logic and scripting). Snakeworlds, finally – a 3D snake game that places the famous classic game on spheres/globes. It is entirely written with Python-Ogre.

WHY LEARN PYTHON?

Okay, that's really where we get into the thick of things. We 're about to begin analyzing many of Python's key features, addressing benefits and drawbacks, and contrasting it to other programming languages. If you are eager for some in-depth knowledge about the features of Python, you've come to the right part.

Let us first take a look at an important yet often misunderstood distinction before we start: coding is not programming. These terms are used interchangeably by most individuals, and it is convenient to do so. However, explaining the concepts behind them can help to create a more beneficial programming attitude amongst beginners.

So, in simple terms, the difference between programming and coding is the same as the difference between creating a story for a novel, and writing it down in a book. Let's do some elaboration. If you want to write a novel, the first step shouldn't be to ask yourself how you would write it in English. Your attention should be on building the story with all the elements thereof. Only then can you prudently write it out in whatever language you want.

For programming, the same principle applies-it is a two-stage process.

First, you 're being faced with a question (say, you need to find the shortest route between cities).

Then you need to come up with steps that will take you to the correct answer in each of those cities' initial graph configurations (we call this an algorithm). This is the most critical programming level and is (mostly) independent of the language in use. Once you have gotten that down, you can start communicating the instructions to your computer through a programming language. That is the stage of codification. Learning code in a particular language by itself is not enough to become a good programmer. When you first study programming, you should concentrate on the methods used to solve a problem, not the language itself.

Having said that, different languages will certainly speed the learning process or slow it down. We think Python is one of the best languages for a beginner to begin programming in. Let's see exactly what:

Python is an open-source language, which means it can be used freely, and everyone can contribute to writing and maintaining their code and libraries. Indeed, many people have devoted time and energy to extending and perfecting Python, and even businesses have. That is a large part of what makes the language so appealing to the community.

Also, Python is a language of high standard and general purpose. High-level implies that it is far from 0s and 1s, and thus closer to the human language. So, you don't have to think about memory management and delete artifacts (among other issues). And that enables you to focus entirely on solving the issue of programming. This makes Python ideal for people who have never previously studied programming as it can greatly enhance their learning experience.

Lastly, general-purpose suggest Python is very versatile. We can use this for software and web creation as well as for network programming, as we described above. On top of this, in particular, data science and machine learning are fields where an application is constantly found. So, it's not shocking that major companies like Facebook, Twitter, Quora, Spotify, and Netflix use Python a lot. Now, although these websites and platforms aren't written in Python, many supporting processes are performed with it, particularly those related to analytics.

Overall, the popularity of Python among developers has continuously increased. The group seems to enjoy it, from the front and back end-users alike. And with its wide range of functionalities, it's not just the major corporations that use it. The IT industry is incorporating Python across the board for the development of different products.

What are the benefits and drawbacks of Python?

So, what are the key aspects of this language of programming which make it so attractive?

Pros:-Pros:

• Easy to understand – simple structure, few keywords, and a syntax that is established. This allows the learner to quickly pick up the language.

• Easy to read – Python code is defined more clearly thanks to the extensive use of white space.

• Productivity – Python code can be written much faster.

• Interactive mode – interactive mode support, which allows interactive testing and debugging of code snippets.

• Broad standard library-Python is known for being the language of "batteries included." There are more than 300 standard library modules, which include classes for a wide range of programming tasks.

• Cross-platform – Python runs equally well on various operating systems such as Windows, Linux, Mac OSX, etc. Consequently, the applications can easily be ported across OS platforms.

• Extendable – The Python Interpreter can be added to low-level modules. These modules allow programmers to add to their tools or customize them to make them more efficient.

• Embeddable-Python can also be embedded. You may put your code into another language's source code, such as C++.

• Difficulty in using other languages – Python followers become so used to its extensive libraries and features that they face problems learning or working with other programming languages. The declaration of variable types and the syntactic requirements of adding curly braces or semicolons can very often be viewed as an onerous task by Python experts.

• Extensive support library collection – numerous Python libraries add a great deal of functionality to the language.

Cons: Counterparts

• Speed limitations-Python is interpreted, resulting in slow code execution. But this is no concern unless pace is the project's main focus. In other words, the benefits Python offers are enough to outweigh its limitations unless high speed is a requirement.

• Weak on mobile – although it has made its appearance on many desktop and server platforms, mobile computing is seen as a poor language. That is the reason it is used by very few mobile applications.

Dynamic, static typing

In describing Python 's characteristics, we cannot fail to note that it is a language typed dynamically. In short, Python variables lack a predefined type (such as an integer or a character string). Instead, the variable type is dynamically determined as the program runs. For example, if we assign value 10 to variable A at one point, then Python will automatically determine that this variable is of type int (integer). If we change it later on to 10.0 it is now going to be afloat (a real number).

Alternatively, the type of variable is written in the code itself in a statically typed language, such as C++. So if you'd like to use A as an integer, write "int A" Now, we can only assign the integer values to A. The program would return an error if you try to assign a non-integer value to A.

Of course, both concepts have their advantages and inconveniences. For a total beginner, it's easier to get into a dynamically typed language. However, since nobody knows the type of variables before runtime, you may get unexpected, hard-to-track errors.

But let's escape these comparisons, somewhat abstract, and see how Python measures against other popular programming languages.

PYTHON OR ANY OTHER LANGUAGE

We compare Python to other programming languages in this section: R, C++, and Java. Once you have read it, you will be able to decide which language of programming best suits your practical needs.

Python is a scripting language. It is very powerful, in the sense that you can execute a wide variety of actions with just a few lines of code. On huge datasets, you can read, visualize, analyze, and even predict, and all that would require just a few lines of code.

R, as it happens, is also a language of the script. So how is it that the two compare? Although both have advantages and disadvantages, Python has been the dominant use of language programmers in the last few years. The reasons why? Well, there are a couple of.

Firstly, although both languages are free and open-source, R is mainly used for statistical analysis (and was developed by statisticians, in fact). Python, on the other hand, is a language of general use (which is an important aspect of language and repeated bears).

General intent means that it is appropriate for all sorts of needs, not just for data science and machine learning, but also for pre-processing, web programming, and just about everything you may think of.

Another value for Python is that it is of high quality. Loosely explained, it has a simple syntax that is similar to rational human language, and is later translated into lower-level (like C) or even 0s and 1s languages. The

NumPy kit, for example, does go through the programming language C. This explains why it is so easy.

Let's notice one last minor thing concerning R. The graphics are not the greatest in R. This is, however, an understatement. By much. The graphics are so bad that R launched R shiny, which was specifically designed to counter the problem. We like Shiny, that's very nice. That said, although in some instances the graphics in Python may also be lacking, on that front, it has recently improved. Seaborn, which is used on top of matplotlib, made the graphics look better than before.

Java & C++ vs Python versus

Now, we stressed the fact that Python is a language of scripts. While that's true, it doesn't give us the complete picture.

While many people use Python mainly to run scripts, it also has object-oriented functionality such as C++ and Java. That said, we find it considerably easier to learn programming with Python than with the other two.

Why, behind the screen, do I hear the seasoned C++ programmer yell? We know you have learned the hard way of programming and are super pleased with it, as it all runs super-fast, and there are no longer as many memory leaks. As long as you know what you do, the code runs, right?

Well, allow us to play the advocate of the devil and consider that the basics of programming could be learned easier and more intuitive.

In the first place, consider Java. For many programmers, the first language.

Java has objects and groups, definitions that are difficult for beginners to grasp. It could entail a long summer of your dad shouting from personal experience, "What's the difference between a class and an object?! You are trying to figure out what the answer is right. The answer for the record is-an object is a class instance. Has this made things clearer? Yeah, that is what we were thinking. Understanding those concepts takes some time. And if you are a beginner, writing and running your code in Java could take you a bit longer.

In a nutshell, that's Java (actually a very small nutshell, but you get the idea).

How about C++, the language of the 'real' programmers?

His motto is "you get what you're paying for" (but on that one, maybe you don't quote us). You are in charge of handling the memory in C++, i.e. creating and removing objects. What's more, the way you access these objects in your memory is by using pointers that can have their own

Take the point(er). C++ is not very friendly to beginners, either.

Let's compare these to programming with Python. But, like, the programming is really simple. (Imagine) you are a complete novice, and you would like to write your first program. This means writing your first "Hello, World" for every programmer, no matter what the language,

So, in Java, the code looks like this:

```java
public class HelloWorld {

    public static void main(String[] args) {
        // Prints "Hello, World" to the terminal window.
        System.out.println("Hello, World");
    }

}
```

And here it is in C++ :

```cpp
#include <iostream>
using namespace std;

int main()
{
    cout << "Hello, World!";
    return 0;
}
```

Now, let's look at Python:

```python
print('Hello World')
```

If the environment is set up and a script file is opened, writing "hello world" in Python needs just a simple print() function and your text. And you don't even need those brackets in Python 2! It doesn't get any simpler than this. This is the readability factor that we were talking about at the start. And that's also the reason we prefer other programming languages to Python. By the way, consider sharing the book with your other coder friends, if you find this analysis useful. One of the reasons it's awesome is the community of Python, so let's help make it bigger.

PYTHON TO LANGUAGES

Python is sometimes contrasted with other languages translated as Java, JavaScript, Perl, Tcl, or Smalltalk. Also, enlightening can be the comparisons to C++, Common Lisp, and Scheme. I will compare Python to each of those languages in this section briefly. Such similarities focus primarily on language issues. In practice, a programming language 's choice is often determined by other real-world constraints such as cost, availability, learning and prior preparation, or even emotional attachment. Since these things are highly complex, much thought for this analogy seems to be a waste of time.

Java

Generally speaking, python programs are expected to run slower than Java programs, but they often take much less time to develop. Python programs are 3-5times shorter than equivalent Java programs. This difference could be attributed to the high-level data types incorporated by Python and their dynamic typing. For instance, a Python programmer wastes no time declaring the types of variables or arguments, and finding use in almost every Python program is the powerful polymorphic list and dictionary types for which rich syntactic support is built directly into the language. Python's runtime will work harder than Java's because of the run-time typing. For instance, when evaluating the expression a+b, the objects a and b must first be inspected to determine their form, which isn't known at compile time. It invokes the appropriate addition operation, which may be a user-defined overloaded method. On the other hand, Java can execute an efficient integer or floating-point addition. However, it requires variable declarations for a and b, and doesn't allow user-defined class instances to overload the + operator.

Python is better suited as a "glue" language for these purposes, while Java is best described as a low-level language for implementation. The two together, in fact, make an excellent blend. Components can be built in Java and combined to shape applications in Python; Python can also be used for prototyping components before a Java implementation can "harden" their design. A Python implementation written in Java is under development to support this form of development, which allows Java to call Python code and vice versa. Python source code is translated into Java byte code in this implementation (with the help of a run-time library to support the dynamic semblance of Python).

Javascript

The "object-based" subset of Python roughly equals JavaScript. Like JavaScript (and unlike Java), Python supports a style of programming that uses basic functions and variables without requiring class definitions. That's all there is for JavaScript, though. On the other hand, Python supports writing much larger programs and better reuse of code through a true object-oriented style of programming, in which classes and inheritance play a key role.

Perl

Python and Perl originate from a similar background (Unix scripting, both long outgrown), and sport a lot of similar features but have a different philosophy. Perl emphasizes support for common application-oriented activities, by having regular expressions built-in, reporting features developed, and file scanning. Python emphasizes support for traditional programming techniques such as object-oriented programming, data structure design, and encourages programmers to write readable and maintainable) code by providing an elegant but not excessively cryptic notation. Consequently, Python comes close to Perl but rarely beats it in its original domain of application; Python, however, has applicability far beyond Perl 's niche.

Tcl Tcl

Like Python, as well as a stand-alone programming language, Tcl is available as an application extension language. However, Tcl is bad on data structures, which typically stores all data as strings, and executes typical code much slower than Python. Tcl also lacks the features required to write large programs, such as modular namespaces. Thus, although a "typical" large application using Tcl typically requires Tcl extensions written in C or C++ specific to that program, it is also possible to write an equivalent Python program in "pure Python." Pure Python development is, of course, much faster than writing and Debug a variable to C or C++. Tcl's one redeeming feature; it has been said, is the Tk toolkit. As their main GUI component library, Python has adopted an interface to Tk.

Tcl 8.0 addresses speed emitters by providing a bytecode compiler with limited support for data type, and adds namespaces. It's a much more cumbersome programming language.

Smalltalk

Maybe the biggest difference between Python and Smalltalk is the more "mainstream" syntax from Python, which gives it a leg up on programmer training. Like Smalltalk, Python has dynamic binding and typing, and it's all an object in Python. Python, however, distinguishes built-in object types from user-defined classes and does not currently allow inheritance from built-in types. The standard library of data collection types for Smalltalk is more refined, while the library for Python has more facilities for dealing with Internet and WWW realities such as email, FTP, and HTML.

Python has a different philosophy regarding code development and distribution. Whereas Smalltalk traditionally has a monolithic "system image" that includes both the environment and the user's program, Python stores standard modules as well as user modules in individual files that can be easily rearranged or distributed outside the system. One consequence of this is there is more than one option to attach a Graphical User Interface (GUI) to a Python program, as the GUI is not integrated into the system.

C+++

Nearly all that has been said about Java also applies to C++, but more so: while Python code is usually 3-5 times shorter than Java code, it is also 5-10 times shorter than C++ code! Anecdotal evidence suggests that one Python programmer can complete what two C++ programmers can't complete in a year in two months. Python shines like a language of the glue used for combining components written in C++.

Common Scheme and Lisp

In their dynamic semantics, these languages are close to Python but so different in their approach to syntax that a comparison becomes a religious argument: is Lisp 's lack of syntax a benefit or a disadvantage? It should be remembered that Python has similar introspective capabilities to Lisp's, and on the fly, Python programmers can create and execute program fragments. Real-world properties are generally decisive: Common Lisp is large (in every sense), and the Scheme world is divided into several incompatible versions, where Python has a single, open, compact implement.

Is Python hard to learn?

The short reply: depends. But the answer you were hoping for is not this, right? Do not worry. Before you start learning Python, we will explain everything you need to know. Beginning with the ...

How long does one have to master Python?

It takes around 3 months of relatively consistent study (based on our estimate) for a complete novice learning how to program well in Python.

But if you plan on using Python for data science, i.e., data analytics or machine learning, the timeframe is usually shorter. Since data science involves a very clear use of the language, understanding the basics will take about a month and two. Full disclosure, we base this on the pace our students complete our training in data science. It takes around 200 hours to complete the 365 Data Science program. It covers the fundamentals of mathematics, statistics, and Python, among other topics. It introduces advanced topics such as the use of Sklearn, pandas, and Numpy for machine learning, and covers deep learning with TensorFlow.

Neural Networking Library. So, if you're truly committed and spend five hours of your day learning, learning the main principles for data science analysis at Python should take you about 1 month.

That said, it depends on ... well, you, the time it takes you to become proficient in Python or any programming language.

Training at different speeds, different individuals. Your ability to learn Python will depend on your programming background, too. If you are already skilled in a programming language, getting familiar with Python will mainly entail getting used to the different libraries.

CHAPTER 6

PYTHON FEATURES

We had about as many programming languages at some point in time as we could count on our fingertips. There are so many here today, and all with their specialties. What makes a language unique, however, are its features. And in the end, it's its features that make it selected or passed for a project. So before we start with deeper Python concepts, let's first look at the basics of Python's programming language, which justifies the reasons behind what makes Python so powerful compared to other programming languages. So let's start with the Python Programing Language Features.

Python Features

1. Easy

We mean it in different contexts when we say the word 'easy.'

a. Simple to code

Python is very easy to code, as we have seen in earlier lessons. Coding in Python is easier compared with other common languages such as C++ and Java. In just a few hours, anyone can learn the Python syntax. Mastering Python, though sure, requires learning about all of its advanced concepts and packages and modules. It takes time. And it's programmer-friendly.

b. Simple to Read

Python programming is somewhat like English, being a high-level language. If you look at it, you can tell what the code should be doing. It also requires indentation, since it is dynamically-typed. This helps in readability.

2. Expressive

Let's first learn about expressiveness. Assume we have two languages A and B, and the local transformations can be used to make all programs that can be made in A. There are, however, some programs that can be made using local transformations in B, but not in A. Then, it says B is more expressive than A. Python offers us a multitude of constructs that help us focus more on the solution than on the syntax. This is one of the outstanding features of python, which tells you why you should learn Python.

3. Free & Open Source

Firstly, Python is available freely. This can be downloaded from the Python Website.

Second, they are open-source. That means the public has access to its source code. You can download them, change them, use them, and distribute them. This is called FLOSS(Free / Free Software and Open Source). As the Python community, we 're all headed towards one goal — a Python that's getting even better.

4. High-Level

As we discussed in point 2b, this is a vocabulary of high quality. This means we do not need to remember the system architecture as programmers. We needn't control the memory either. That makes it more programmer-friendly and is one of the key features of python.

5. Portable

Let's say you wrote a Python file for your Windows computer. But if you want to run it on a Pc, you don't have to make the same improvements to it. In other words, you can take 1 code and run it on any machine; for different machines, there is no need to write different code. This makes for a compact language for Python. In this case, however, you need to avoid any system-dependent features.

6. Interpreted Version

If you are familiar with any languages such as Java or C++, you need to compile it first, then run it. But it doesn't need to be compiled at Python. Internally, it transforms the source code into an immediate form called bytecode. So, just running your Python code without thinking about connecting to libraries and a few other things is all you need to do.

By reading, we say that the source code is run line by line and not all at once. Because of this, debugging the code is simpler. Interpreting also makes it slightly slower than Java, but that doesn't matter when compared with the advantages that it has to bring.

7. Goal-focused

It is said that it is an object-oriented programming language that can model the real world. It focuses on objects and puts together data and functions. Contrary to this, a language geared towards procedures revolves around functions, which are code that can be reused. Python supports both procedure-oriented programming and object-oriented programming, which is one of the key features of python. In contrast to Java, it also supports multiple inheritances. A class is a blueprint for an object like this. It is an abstract type of data and contains no values.

8. Enlarge

You may write some of your Python code in other languages, such as C++, if necessary. This makes Python an extensible language, which means it can be extended to include other languages.

240 + Python Tutorials – Master Python programming in real-time and practical projects.

9. Embedded

We've just seen that we can insert code into our Python source code in other languages. However, you can also bring our Python code into source code in another language like C++. This enables us to integrate the scripting capabilities into the other language program.

10. Large Standard Library

Python downloads that you can use with a large library, so you don't have to write your code for every single thing. There are libraries for regular expressions, generation of documents, unit testing, web browsers, threading, databases, CGI, email, image manipulation, and many other features.

11. Programmable GUI

Until their GUI is made, a software is not user-friendly. A user can interact easily with a GUI to the software. Python provides a range of libraries to render Graphical UI for your applications. You may make use of Tkinter, wxPython, or JPython for this. These toolkits allow you to build GUI easily and quickly.

12. Typed Dynamically

Python is typed dynamically. This means that you decide the type for a value at runtime, not beforehand. Therefore we do not need to specify the data type while declaring it.

The python programming language tutorial features are all about this.

THE BENEFITS OF PYTHON LEARNING

There are many advantages to studying and working with Python.

For example, learning is extremely simple and can be used as a move in other languages and frameworks such as PERL, C, C++, and more. If you are an absolute beginner and this is your first time working with any kind of coding language, that is certainly something you want. Once you complete your training, you'll have plenty of opportunities to grow if necessary, rather than being confined to one language.

Python is then popular because it is commonly used. So famous, that several tech giants like Instagram, Google, Pinterest, Yahoo use this! – IBM, Disney, Nokia, etc. Once you've learned Python, there'll never be a shortage of ways to use that skill. You can make good cash as a Python developer because a lot of big companies rely on the language.

Many advantages include:

1) Python can be used in product production and can help speed up the idea of the design process as it is so easy to use and to understand.

2) Python is ideally suited for general purpose tasks such as data mining and the facilitation of big data.

3) Developers of all ability levels prefer to remain more coordinated and efficient as compared to languages like C # and Java as operating with Python.

4) Python is easy to read, even if you are not an experienced programmer, so it is ideal for use among multi-programmers and large development teams, especially those with inexperienced coding team members.

5) Django is a comprehensive and open-source Web application framework powered by Python. Frameworks – like Ruby on Rails – simplify the process of development by allowing developers to work with existing code snippets called modules. These packets of code can be modified and repurposed over multiple projects as necessary.

6) Because Python is an open-source language and a developed community, it has a massive base of support. Millions of like-minded developers frequently interact with the language. Additionally, the community works together continuously to improve core functionality. This is also a perfect place for other developers to network.

CHAPTER 7

ANALYSIS USING PANDAS IN PYTHON

Let me introduce you to another animal (as if Python wasn't enough!) - Pandas to further explore our results.

Pandas is one of Python's most useful data analysis library (I know those names sound strange but hang on!). They played a crucial role in increasing the use of Python in the data science community. We'll now use Pandas to read a data set from competition for Analytics Vidhya, conduct exploratory analysis, and build our first basic categorization algorithm to solve this problem.

Let's understand the two key data structures in Pandas – Series, and DataFrames before loading the data

INTRODUCTION TO DATA FRAMES AND SERIES

Series can be interpreted as a labeled / indexed sequence of 1 dimension. These labels allow you to access individual elements of this series.

A data frame is similar to Excel workbook – you have column names that apply to columns and rows that can be accessed using row numbers. The essential difference is that, in the case of data frames, column names and row numbers are known as column and row index.

Series and data frames form Pandas in Python 's core data model. First, the data sets are to be read into these data frames, and then various operations (e.g., group by, grouping, etc.) can be applied to their columns very quickly.

Practice data set – Loan Prediction Problem

VARIABLE   DESCRIPTIONS:

Variable	Description
Loan_ID	Unique Loan ID
Gender	Male/ Female
Self_Employed	Self-employed (Y/N)
Dependents	Number of dependents
Married	Applicant married (Y/N)
Education	Applicant Education (Graduate/ Under Graduate)
ApplicantIncome	Applicant income
CoapplicantIncome	Coapplicant income
LoanAmount	Loan amount in thousands
Loan_Amount_Term	Term of loan in months
Credit_History	credit history meets guidelines
Property_Area	Urban/ Semi-Urban/ Rural
Loan_Status	Loan approved (Y/N)

Let's start with investigation

Start iPython GUI in Inline Pylab mode by typing to your terminal/windows command prompt as follows:

```
python notebook --pylab=inline
```

This opens up an iPython notebook in pylab environment, which has already imported a few useful libraries. You will also be able to map the data inline, which makes this a very nice interactive data analysis environment. By typing the following command (and getting the output as shown in the figure below), you can check whether the environment has loaded correctly:

plot(arrange(5))

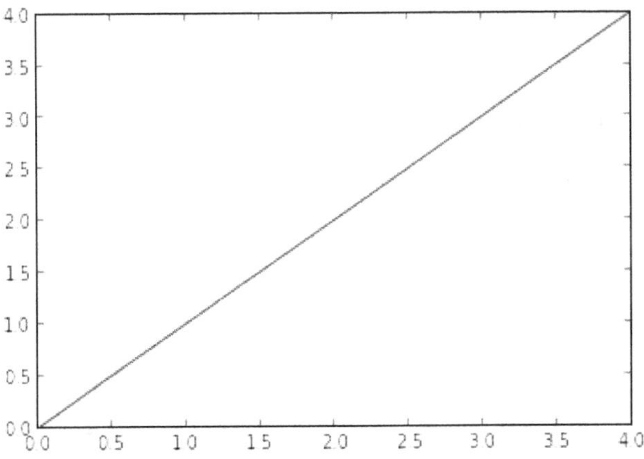

Libraries Import and Data Set:

The libraries which we will use during this guide will follow:

• Matplotlib

• Numpy

• To pandas

Please note that due to the Pylab environment, you don't need to import matplotlib and NumPy. In case you use the code in a different environment, I still kept them in the code.

Using read CSV) (function, you read the dataset after importing the library. So until this point, the code looks like this:

import pandas as PD

import NumPy as np

import matplotlib as plt

%matplotlib inline

df = PD.read_csv("/home/kunal/Downloads/Loan_Prediction /train.csv") #Reading the dataset in a data frame using Pandas

Quick Data Exploration

Once you have read the dataset, use the head () feature to look at a few top rows

df.head(10)

In [3]: df.head(10) #Printing first 10 rows of dataset

Out[3]:

	Loan_ID	Gender	Married	Dependents	Education	Self_Employed	ApplicantIncome	CoapplicantIncome	LoanAmount	Loan_Amount_Term	Cr
0	LP001002	Male	No	0	Graduate	No	5849	0	NaN	360	1
1	LP001003	Male	Yes	1	Graduate	No	4583	1508	128	360	1
2	LP001005	Male	Yes	0	Graduate	Yes	3000	0	66	360	1
3	LP001006	Male	Yes	0	Not Graduate	No	2583	2358	120	360	1
4	LP001008	Male	No	0	Graduate	No	6000	0	141	360	1
5	LP001011	Male	Yes	2	Graduate	Yes	5417	4196	267	360	1
6	LP001013	Male	Yes	0	Not Graduate	No	2333	1516	95	360	1
7	LP001014	Male	Yes	3+	Graduate	No	3036	2504	158	360	0
8	LP001018	Male	Yes	2	Graduate	No	4006	1526	168	360	1
9	LP001020	Male	Yes	1	Graduate	No	12841	10968	349	360	1

This should take 10 rows to print. Alternatively, by printing the file, you can look at more rows too.

Next, you can look at summaries of numerical fields using the function describe()

df.describe()

In [4]: df.describe() #Get summary of numerical variables

Out[4]:

	ApplicantIncome	CoapplicantIncome	LoanAmount	Loan_Amount_Term	Credit_History
count	614.000000	614.000000	592.000000	600.00000	564.000000
mean	5403.459283	1621.245798	146.412162	342.00000	0.842199
std	6109.041673	2926.248369	85.587325	65.12041	0.364878
min	150.000000	0.000000	9.000000	12.00000	0.000000
25%	2877.500000	0.000000	100.000000	360.00000	1.000000
50%	3812.500000	1188.500000	128.000000	360.00000	1.000000
75%	5795.000000	2297.250000	168.000000	360.00000	1.000000
max	81000.000000	41667.000000	700.000000	480.00000	1.000000

The describe () function will include the count, mean, standard deviation (std), min, quartile and max in its output (Read this guide to refresh simple statistics to understand population distribution)

Here are a few inferences which you can draw by looking at the describe) (function output:

1. LoanAmount has 22 missing values (614-592).

2. Loan Amount Term has 14 missing values (614-600).

3. Credit History has 50 missing values (614-564).

4. We should also look at why there is a credit history for around 84 percent of applicants. How? How? The mean of the Credit History area is 0.84 (Remember, for those with a credit history, Credit History has meaning 1, and 0 otherwise)

The distribution of Applicant Income would appear to be in line with expectations. The same is true of Co-applicant Income

Please note that by comparing the mean to the median, i.e., the 50 percent figure, we can get an idea of a possible skew in the data.

For the non-numeric values (e.g., Property Area, Credit History, etc.), we should look at the distribution of frequencies to understand how they make sense. The frequency table is printable using the following command:

df['Property_Area'].value_counts()

Similarly, we can look at unique credit history port values. Note that dfname['column name] 'is a basic indexing technique for accessing a given data frame column. It can also be a column-list. For more information, refer to the above-shared resource "10 Minutes to Pandas."

Distribution analysis

Now that we know the basic data characteristics, let's research the distribution of different variables. Start with numerical variables-ApplicantIncome and LoanAmount

Let Applicant-Income start by plotting the histogram using the following commands:

df['ApplicantIncome'].hist(bins=50)

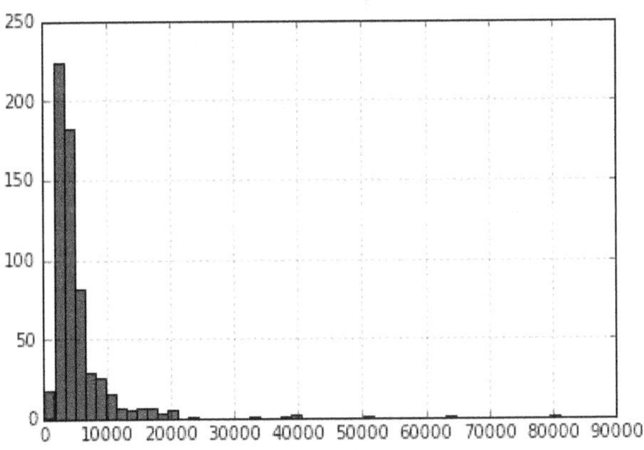

Here we note that the extreme values are low. This is also the reason why it takes 50 bins to represent the delivery.

Next, to understand the distributions, we look at box plots. Farebox plot can be plotted using:

df.boxplot(column='ApplicantIncome')

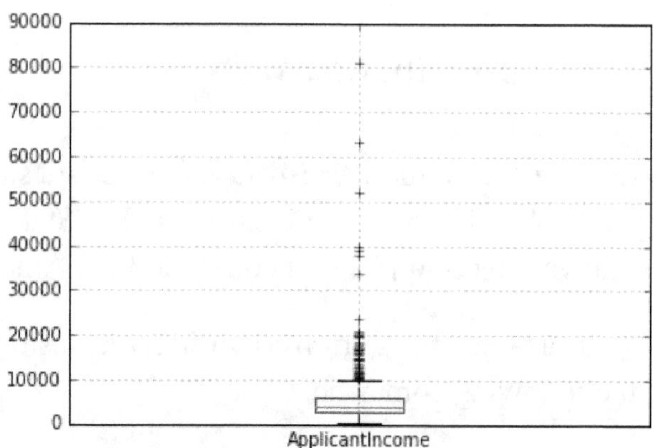

This shows that a lot of outliers / extreme values are present. That can be attributed to the societal income disparity. Part of this may be motivated by the fact we look at individuals of different levels of education. Separate them by Education:

df.boxplot(column='ApplicantIncome', by = 'Education')

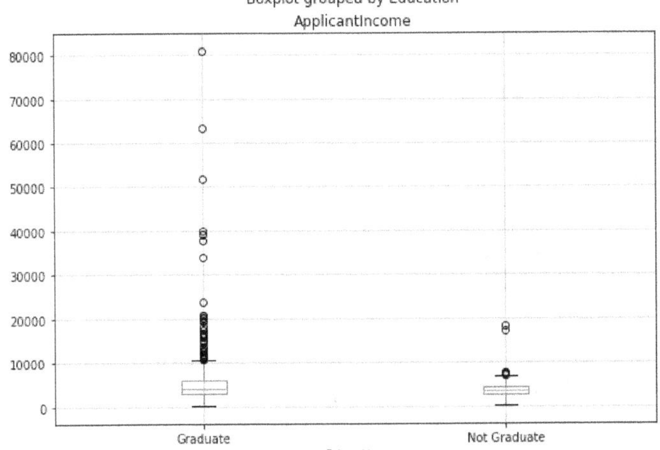

We can see that there is no major difference between the graduate and non-graduate mean incomes. But there are more graduates with very high incomes, which appear to be the outliers.

Let's now take a look at the LoanAmount histogram and boxplot using the following command:

```
df['LoanAmount'].hist(bins=50)
```

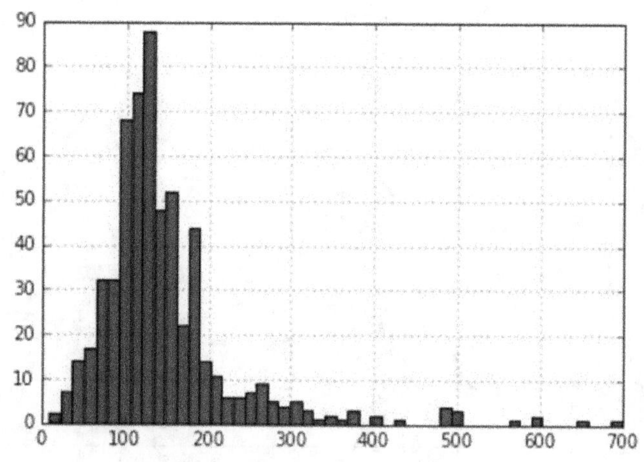

df.boxplot(column='LoanAmount')

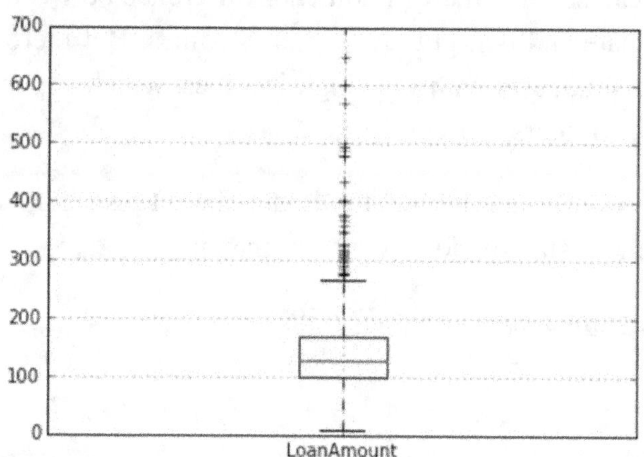

Again, some extreme values do exist. Both ApplicantIncome and LoanAmount require some munging of the data. LoanAmount has missing and extreme values, while ApplicantIncome has some extreme values that need a deeper understanding. We 're going to take that up in sections to come.

Analysis of Categorical Variables

Now that we understand the Loan-Income and Applicant-Income distributions, let 's explain categorical variables in more detail. We'll use a pivot table and cross-tabulation in Excel format. Let's look at the probability of securing a loan based on the credit history, for example. This can be done with a pivot table in MS Excel as:

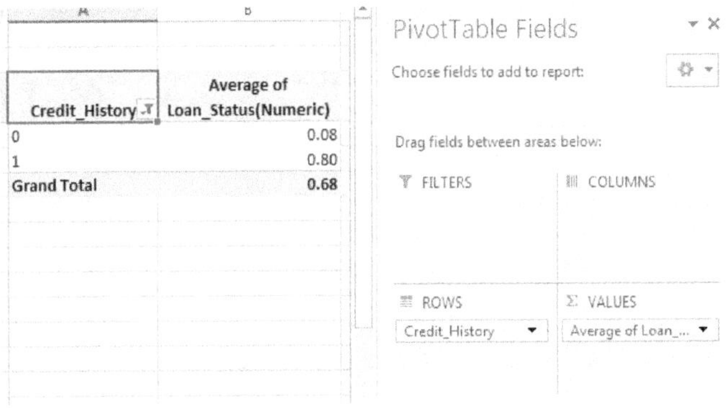

Note: The loan status here was coded as 1 for Yes and 0 for No. So the mean represents the likelihood of obtaining a loan.

Now we're going to look at the steps needed to produce a similar insight using Python.

```
temp1 = df['Credit_History'].value_counts(ascending=True)
```

temp2 =
df.pivot_table(values='Loan_Status',index=['Credit_Hist
ory'],aggfunc=lambda x: x.map({'Y':1,'N':0}).mean())

print ('Frequency Table for Credit History:')

print (temp1)

print ('\nProbility of getting loan for each Credit History
class:')

print (temp2)

```
Frequency Table for Credit History:
0      89
1     475
Name: Credit_History, dtype: int64

Probility of getting loan for each Credit History class:
Credit_History
0     0.078652
1     0.795789
Name: Loan_Status, dtype: float64
```

Now we can observe that we have a pivot table similar to
the MS Excel one. This can be plotted as a bar chart using
a library called "matplotlib" with the following code:

import matplotlib.pyplot as plt

fig = plt.figure(figsize=(8,4))

ax1 = fig.add_subplot(121)

ax1.set_xlabel('Credit_History')

ax1.set_ylabel('Count of Applicants')

ax1.set_title("Applicants by Credit_History")

temp1.plot(kind='bar')

ax2 = fig.add_subplot(122)

temp2.plot(kind = 'bar')

ax2.set_xlabel('Credit_History')

ax2.set_ylabel('Probability of getting loan')

ax2.set_title("Probability of getting loan by credit history")

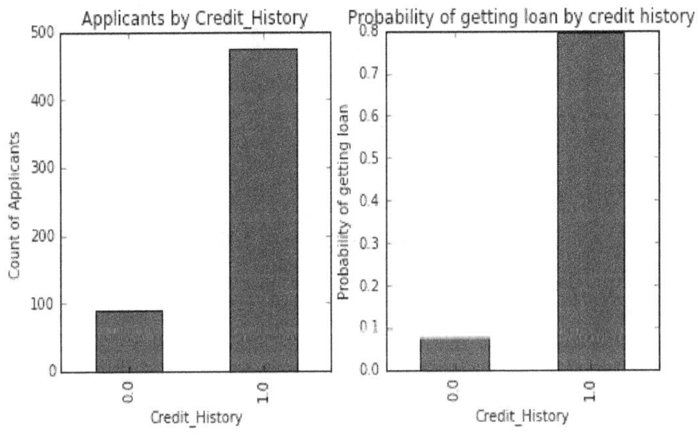

This shows that if the applicant has a valid credit history, the chances of obtaining a loan will be eight times that. You can use Married, Self-Employed, Property Area to plot similar graphs, etc.

Additionally you can also represent these two plots by merging them in a stacked chart:

temp3 = pd.crosstab(df['Credit_History'], df['Loan_Status'])

temp3.plot(kind='bar', stacked=True, color=['red','blue'], grid=False)

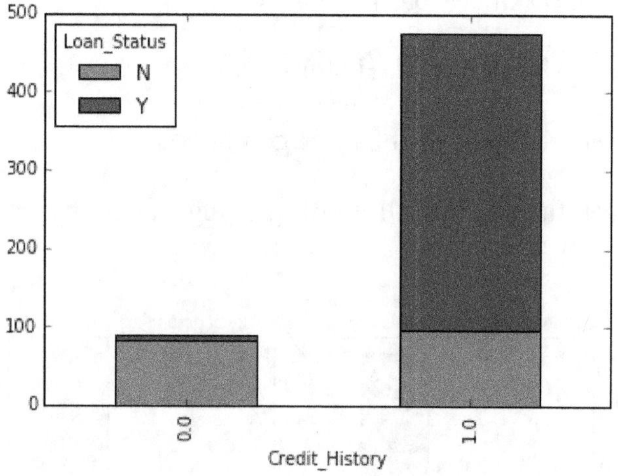

You can also add gender to the mix (similar to the Excel pivot table):

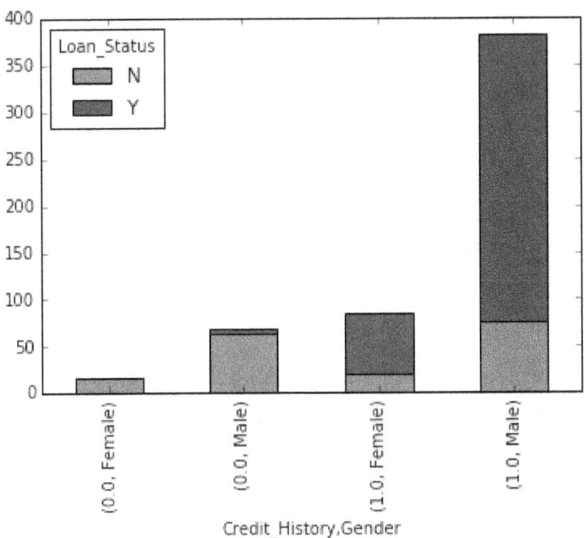

If you haven't already realized that, here we've just created two basic classification algorithms, one based on credit history, while another one based on 2 categorical variables (including sex). To build your first submission on AV Datahacks, you can code this easily.

We just saw how we could use Pandas to do exploratory analyzes in Python. I hope your passion for pandas (the animal) must have increased by now – the library will provide you with the amount of support in analyzing datasets.

Next, let's further explore the variables of ApplicantIncome and LoanStatus, perform data munging, and create a dataset to apply different modeling techniques. I would urge you to take a different dataset and problem and go through an independent example before further reading.

VARIABLES IN PYTHON

This lesson focuses on variables. Many who already know some programming ought to know about the definition of variable and its meaning. Variable are nameplates that we place on empty boxes (memory locations) where we can store any value(data), and when we use a variable, the nameplate is removed from the box(memory location), which can then have a new nameplate. In python, one box may also have multiple names.

All is an object in python, and variables are just names provided to define certain objects.

For instance:

```
x = 10
```

In the code above, x is the object 10 variable or label, or name.

If the code we have is as follows:

```
x = 10
y = 10
```

Then, we have two labels for object 10 (references), x, and y.

Therefore, we can say that variables provide a means to label data with a descriptive name so that the reader and ourselves can understand our programs more clearly. It is useful to think about variables as nameplates for containers holding information. Its sole purpose is to label the data that is stored in memory. You can then use this data throughout your program.

Let's see just a few examples. A feature will have a name on it. The variable has certain rules for assigning a name.

• The name of the variable can consist only of the number(s), alphabet(s), and underscore(s).

• A number can not be the first character in the name of a variable. Consequently, I variable! All invalid variable names are, 1variable, # variable. Although the variable I am variable is all true names, variable 1, variable is.

Save value in variable

Time to find out how to store a value in our variable. Consider a variable called x, we want that variable to store a number or a numeric value, say 11. Then, just go to IDLE and type to do this:

>>> x = 11

And then press Enter. Thereby our variable is created, named x, and stored in it with default value 11, using the operator equal to =. Remember that Equal to operator is always used to assign any variable to a specific value. The name of the variable will always be on the left side and the value on the right. Let's build a different attribute, say y, and assign value 25.

```
>>> y = 25
```

Python is now dealing with two variables in the current IDLE session, x and y, which have values 11 and 25 respectively. If you want to check any variable's value in IDLE, simply type that variable 's name into a new line and press the Enter key. The value that is stored in it will be written in a new line on the IDLE panel.

```
>>> x

11

>>> y

25
```

Now try to look at the below code, what do you think this will do?

```
>>> x = y
```

As you can see on the LHS(Left Hand Side), we have x and y on the RHS(Right Hand Side). Thus the value on the right will be assigned to the variable on the left, as stated earlier. Since y has a value of 25 stored in it, this statement will alter the value from 11 to 25 inside x. And therefore, if you ask for the value of x again, it will now be 25. In this case, the value inside x variable we just overwrote.

>>> x

25

```
>>> x = 11
>>> y = 25
>>> x
11
>>> y
25
>>> x=y
>>> x
25
>>> y
25
>>>
```

Once you're done with the example above, this time, let's be more creative while naming our variables. Let's create a variable and assign it as its value to your name. So here we have a box named name, that can store any word. The method is pretty similar to what we've done above but with just a slight adjustment. Careful watch,

>>> name = "Sudytonight"

As you can see, within the double quotation marks we cited the name of our website. This is because we don't want to get confused about python compilers. Since Studytonight is a term (or, more specifically, string in the world of programming), we'll have to surround it with quotation marks. In doing so, we 're telling python it's a word. But, what should happen if we write Studytonight without the quote marks? Such is the case with

>>> name = Studytonight

Since there are no quotation marks, python will consider Studytonight as another variable and will try to find the stored value within it so that the variable name can be further assigned to it. But since we've never declared any variable with the name Studytonight, python won't be able to find any value for it, and ultimately, it will throw an error saying the variable with the name Studytonight isn't set.

Also, you can use both single quotation and double quotation to describe a word (or string).

>>> name = "Studytonight.com"

>>> name = 'Studytonight'

<u>Live Sample</u> →

Both are fine.

```
>>> name = "Saharsh"
>>> name = 'Saharsh'
>>> name
'Saharsh'
>>> name = Saharsh

Traceback (most recent call last):
  File "<pyshell#86>", line 1, in <module>
    name = Saharsh
NameError: name 'Saharsh' is not defined
>>>
```

Now, as we learned in the previous guide, you can try using variables with math functions too. Seek to use such variables as the function arguments. For instance,

162

```
>>> x = 3
>>> y = 2
>>> pow(x, y)
9
>>> z = -7
>>> abs(z)
7
```

Next, you can try to save the answer in a variable to any math function. Such as,

```
>>> p = pow(5, 2)
>>> import math
>>> q = math.log(x)
```

Seek using certain variables with mathematical operators. Such as,

```
>>> x+y
5
>>> x-y
1
```

Seek to use these variables in a mathematical expression that includes some mathematical function and operators, like,

>>> x = pow(x**y, p+2)

In the code above, python first calculates the expression on the RHS(Right Hand Side) and uses the old value for variable x, which is 3, and once the expression is resolved, the answer will be saved in variable x, which becomes its new value.

```
>>> x = 3
>>> y = 2
>>> pow(x, y)
9
>>> z = -7
>>> abs(z)
7
>>> p = pow(5, 2)
>>> import math
>>> q = math.log(x)
>>> p
25
>>> q
1.0986122886681098
>>> x+y
5
>>> x-y
1
>>> x = pow(x**y, p+2)
>>> x
58149737003040059690390169L
>>>
```

CHAPTER 8

PYTHON MACHINE LEARNING ALGORITHMS YOU MUST LEARN

Objective

We've discussed the machine learning techniques with Python before. Going deeper, we are studying and implementing 8 top Machine Learning Algorithms in Python today.

In Python Programming, let 's continue the journey of Machine Learning Algorithms.

Python, Machine learning algorithms

Following are the Python Machine Learning Algorithms:

a. Linear Regression

Linear regression is one of Python's supervised Machine learning algorithms which observes continuous characteristics and predicts an outcome. We may call it simple linear regression, or multiple linear regression, depending on whether it operates on a single variable or several apps.

This is one of the most popular, and often under-appreciated Python ML algorithms. To create a line ax+b to predict the output, it assigns optimum weights to variables. Often we use linear regression to estimate real values such as several house calls and costs based on continuous variables.

The regression line is the best line to denote a relationship between independent and dependent variables that matches $Y = a*X+b$.

Do you know about configuration of Python Machine Learning System

Let's plot that for the dataset on diabetes.

```
>>> import matplotlib.pyplot as plt
```

```
>>> import numpy as np
```

```
>>> from sklearn import datasets,linear_model
```

```
>>> from sklearn.metrics import mean_squared_error,r2_score
```

```
>>> diabetes=datasets.load_diabetes()
```

```
>>> diabetes_X=diabetes.data[:,np.newaxis,2]
```

```
>>> diabetes_X_train=diabetes_X[:-30] #splitting data
into training and test sets

>>> diabetes_X_test=diabetes_X[-30:]

>>> diabetes_y_train=diabetes.target[:-30]   #splitting
targets into training and test sets

>>> diabetes_y_test=diabetes.target[-30:]

>>> regr=linear_model.LinearRegression()   #Linear
regression object

>>> regr.fit(diabetes_X_train,diabetes_y_train)   #Use
training sets to train the model

LinearRegression(copy_X=True,      fit_intercept=True,
n_jobs=1, normalize=False)

>>>      diabetes_y_pred=regr.predict(diabetes_X_test)
#Make predictions

>>> regr.coef

array([941.43097333])

>>>
mean_squared_error(diabetes_y_test,diabetes_y_pred)

3035.0601152912695

>>>      r2_score(diabetes_y_test,diabetes_y_pred)
#Variance score
```

0.410920728135835

```
>>>    plt.scatter(diabetes_X_test,diabetes_y_test,color
='lavender')
```

<matplotlib.collections.PathCollection   object   at
0x0584FF70>

```
>>>
plt.plot(diabetes_X_test,diabetes_y_pred,color='pink',li
newidth=3)
```

[<matplotlib.lines.Line2D object at 0x0584FF30>]

```
>>> plt.xticks(())
```

([], <a list of 0 Text xticklabel objects>)

```
>>> plt.yticks(())
```

([], <a list of 0 Text yticklabel objects>)

```
>>> plt.show()
```

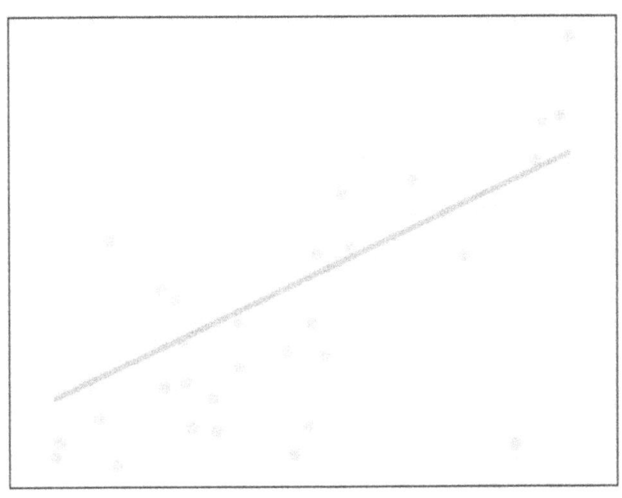

Python Machine Learning Algorithms-Linear Regression

Python Machine Learning – Preprocessing, Analysis and Visualisation of data

b. Logistic loss

Logistic regression is a supervised classification to unique Python Machine Learning algorithms used to estimate discrete values like 0/1, yes / no, and true/false. This is based on an unrelated set of variables. We use a logistic function to predict the likelihood of an event, and this gives us an output of 0 to 1.

Even though it says 'regression,' it is an algorithm for classification. Logistic regression fits data into a logit function and is also known as regression logit. Let's just plot it.

```
>>> import numpy as np

>>> import matplotlib.pyplot as plt

>>> from sklearn import linear_model

>>> xmin,xmax=-7,7  #Test set; straight line with Gaussian noise

>>> n_samples=77

>>> np.random.seed(0)

>>> x=np.random.normal(size=n_samples)

>>> y=(x>0).astype(np.float)

>>> x[x>0]*=3

>>> x+=.4*np.random.normal(size=n_samples)

>>> x=x[:,np.newaxis]
```

```
>>>         clf=linear_model.LogisticRegression(C=1e4) #Classifier

>>> clf.fit(x,y)

>>> plt.figure(1,figsize=(3,4))

<Figure size 300x400 with 0 Axes>

>>> plt.clf()

>>> plt.scatter(x.ravel(),y,color='lavender',zorder=17)

<matplotlib.collections.PathCollection object at 0x057B0E10>

>>> x_test=np.linspace(-7,7,277)

>>> def model(x):

return 1/(1+np.exp(-x))

>>> loss=model(x_test*clf.coef_+clf.intercept_).ravel()

>>> plt.plot(x_test,loss,color='pink',linewidth=2.5)

[<matplotlib.lines.Line2D object at 0x057BA090>]

>>> ols=linear_model.LinearRegression()

>>> ols.fit(x,y)

LinearRegression(copy_X=True,     fit_intercept=True, n_jobs=1, normalize=False)
```

```
>>> plt.plot(x_test,ols.coef_*x_test+ols.intercept_,linewidth
=1)

[<matplotlib.lines.Line2D object at 0x057BA0B0>]

>>> plt.axhline(.4,color='.4')

<matplotlib.lines.Line2D object at 0x05860E70>

>>> plt.ylabel('y')

Text(0,0.5,'y')

>>> plt.xlabel('x')

Text(0.5,0,'x')

>>> plt.xticks(range(-7,7))

>>> plt.yticks([0,0.4,1])

>>> plt.ylim(-.25,1.25)

(-0.25, 1.25)

>>> plt.xlim(-4,10)

(-4, 10)

>>> plt.legend(('Logistic Regression','Linear
Regression'),loc='lower right',fontsize='small')

<matplotlib.legend.Legend object at 0x057C89F0>
```

Do you know How to Split Train and Test Set in Python Machine Learning

1.   >>> plt.show()

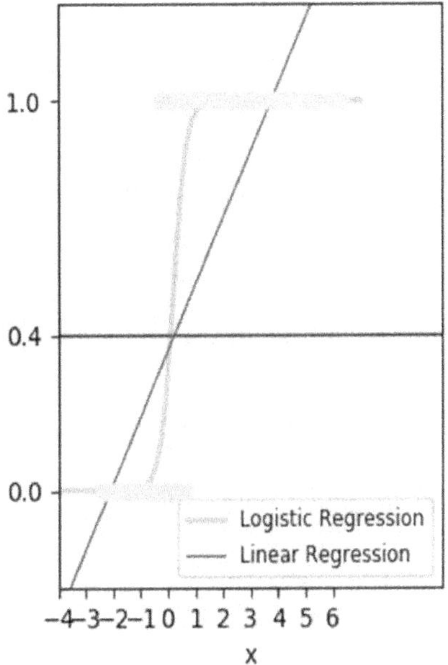

Machine Learning Algorithms in Python -Logistic

c. Arbitrary decision

A decision tree falls under Python's supervised Machine Learning Algorithms and is used for classification as well as regression-though mostly for classification. This example takes an instant, passes through the tree, and compares important features with a conditional statement determined. If it descends to the branch of the left child or the right depends on the result. More significant features are typically more similar to the surface.

Decision Tree, a Python Machine Learning algorithm can work on both categorical and continuous variables dependent upon it. Here, we are dividing a population into two or more homogenous sets. Let's look at the algorithm here-

```python
>>> from sklearn.cross_validation import train_test_split

>>> from sklearn.tree import DecisionTreeClassifier

>>> from sklearn.metrics import accuracy_score

>>> from sklearn.metrics import classification_report

>>> def importdata(): #Importing data

balance_data=pd.read_csv('https://archive.ics.uci.edu/ml/machine-learning-'+

'databases/balance-scale/balance-scale.data',

sep= ',', header = None)

print(len(balance_data))

print(balance_data.shape)

print(balance_data.head())

return balance_data

>>> def splitdataset(balance_data): #Splitting data
```

```python
x=balance_data.values[:,1:5]

y=balance_data.values[:,0]

x_train,x_test,y_train,y_test=train_test_split(

x,y,test_size=0.3,random_state=100)

return x,y,x_train,x_test,y_train,y_test
>>>    def    train_using_gini(x_train,x_test,y_train): #Training with giniIndex

clf_gini = DecisionTreeClassifier(criterion = "gini",

random_state = 100,max_depth=3, min_samples_leaf=5)

clf_gini.fit(x_train,y_train)

return clf_gini
>>>   def train_using_entropy(x_train,x_test,y_train): #Training with entropy

clf_entropy=DecisionTreeClassifier(

criterion = "entropy", random_state = 100,

max_depth = 3, min_samples_leaf = 5)

clf_entropy.fit(x_train,y_train)
return clf_entropy
>>>    def    prediction(x_test,clf_object):    #Making predictions
```

```python
y_pred=clf_object.predict(x_test)

print(f"Predicted values: {y_pred}")

return y_pred

>>> def cal_accuracy(y_test,y_pred):  #Calculating accuracy

print(confusion_matrix(y_test,y_pred))

print(accuracy_score(y_test,y_pred)*100)

print(classification_report(y_test,y_pred))

>>> data=importdata()

625
(625, 5)

  0 1 2 3 4

0 B 1 1 1 1

1 R 1 1 1 2

2 R 1 1 1 3

3 R 1 1 1 4

4 R 1 1 1 5

>>> x,y,x_train,x_test,y_train,y_test=splitdataset(data)

>>> clf_gini=train_using_gini(x_train,x_test,y_train)
```

>>>

clf_entropy=**train_using_entropy**(x_train,x_test,y_train
)

>>> y_pred_gini=**prediction**(x_test,clf_gini)

```
>>> y_pred_gini=prediction(x_test,clf_gini)
Predicted values: ['R' 'L' 'R' 'R' 'R' 'L' 'R' 'L' 'L' 'L' 'R' 'L' 'L' 'L' 'R' 'L' 'R' 'L'
 'L' 'R' 'L' 'R' 'L' 'L' 'R' 'L' 'L' 'L' 'R' 'L' 'L' 'L' 'R' 'L' 'L' 'L'
 'L' 'R' 'L' 'L' 'R' 'L' 'R' 'L' 'R' 'R' 'L' 'L' 'R' 'L' 'R' 'R' 'L' 'R'
 'R' 'L' 'R' 'R' 'L' 'L' 'R' 'R' 'L' 'L' 'L' 'L' 'L' 'R' 'R' 'L' 'L' 'R'
 'R' 'L' 'R' 'L' 'R' 'R' 'R' 'L' 'R' 'L' 'L' 'L' 'L' 'R' 'R' 'L' 'R' 'L'
 'R' 'R' 'L' 'L' 'L' 'R' 'R' 'L' 'L' 'L' 'R' 'L' 'R' 'R' 'R' 'R' 'R'
 'R' 'L' 'R' 'L' 'R' 'R' 'L' 'R' 'R' 'R' 'R' 'R' 'L' 'R' 'L' 'L' 'L'
 'L' 'L' 'L' 'R' 'R' 'R' 'R' 'L' 'R' 'R' 'R' 'L' 'L' 'R' 'L' 'R' 'L'
 'L' 'L' 'R' 'L' 'L' 'R' 'L' 'R' 'L' 'R' 'R' 'R' 'L' 'R' 'R' 'R' 'R'
 'L' 'L' 'R' 'R' 'R' 'R' 'L' 'R' 'R' 'R' 'L' 'R' 'L' 'L' 'L' 'L' 'R'
 'L' 'R' 'R' 'L' 'L' 'R' 'R' 'R']
```

>>> **cal_accuracy**(y_test,y_pred_gini)

[[0 6 7]

[0 67 18]

[0 19 71]]

73.40425531914893

```
              precision    recall  f1-score   support

           B       0.00      0.00      0.00        13
           L       0.73      0.79      0.76        85
           R       0.74      0.79      0.76        90

 avg / total       0.68      0.73      0.71       188
```

Machine Learning Algorithms in Python – Decision Tree
>>> y_pred_entropy=**prediction**(x_test,clf_entropy)

```
>>> y_pred_entropy=prediction(x_test,clf_entropy)
Predicted values: ['R' 'L' 'R' 'L' 'R' 'L' 'R' 'L' 'R' 'R' 'R' 'R' 'L' 'L' 'R' 'L' 'R' 'L'
 'L' 'R' 'L' 'R' 'L' 'L' 'R' 'L' 'R' 'L' 'R' 'L' 'R' 'L' 'R' 'L' 'L' 'L'
 'L' 'L' 'R' 'L' 'R' 'L' 'R' 'L' 'R' 'R' 'L' 'L' 'R' 'L' 'L' 'R' 'L' 'L'
 'R' 'L' 'R' 'R' 'L' 'R' 'R' 'R' 'L' 'L' 'R' 'L' 'L' 'R' 'L' 'L' 'L' 'R'
 'R' 'L' 'R' 'L' 'R' 'R' 'R' 'L' 'R' 'L' 'L' 'L' 'L' 'R' 'R' 'L' 'R' 'L'
 'R' 'R' 'L' 'L' 'L' 'R' 'R' 'L' 'L' 'L' 'R' 'L' 'L' 'R' 'R' 'R' 'R' 'R'
 'R' 'L' 'R' 'L' 'R' 'R' 'L' 'R' 'R' 'L' 'R' 'R' 'L' 'R' 'R' 'R' 'L' 'L'
 'L' 'L' 'L' 'R' 'R' 'R' 'R' 'L' 'R' 'R' 'R' 'L' 'L' 'R' 'L' 'R' 'L' 'R'
 'L' 'R' 'R' 'L' 'L' 'R' 'L' 'R' 'R' 'R' 'R' 'R' 'L' 'R' 'R' 'R' 'R' 'R'
 'R' 'L' 'R' 'L' 'R' 'R' 'L' 'R' 'L' 'R' 'L' 'R' 'L' 'L' 'L' 'L' 'L' 'R'
 'R' 'R' 'L' 'L' 'L' 'R' 'R' 'R']
```

Machine Learning Algorithms in Python – Decision Tree

>>> **cal_accuracy**(y_test,y_pred_entropy)

[[0 6 7]

[0 63 22]

[0 20 70]]

70.74468085106383


```
                 precision    recall  f1-score   support

             B       0.00      0.00      0.00        13
             L       0.71      0.74      0.72        85
             R       0.71      0.78      0.74        90

    avg / total       0.66      0.71      0.68       188
```

Machine Learning Algorithms in Python – Decision Tree
Algorithm i

With Python, let 's explore 4 machine learning techniques

d. Assisting Vector Machines (SVM)

SVM is a supervised classification, one of Python's most important Machine Learning algorithms, which plots a line dividing your data into different categories. We measure the vector in this ML algorithm to optimize the rows. It is to ensure that any group's closest point lies the farthest from each other. While you'll almost always find that this is a linear vector, it can be different from that.

In this tutorial about Python Machine Learning, we plot each item of data as a point in an n-dimensional space. We have n traits and each trait has a certain coordinate value.

```
>>> from sklearn.datasets.samples_generator import make_blobs

>>> x,y=make_blobs(n_samples=500,centers=2,

random_state=0,cluster_std=0.40)

>>> import matplotlib.pyplot as plt

>>> plt.scatter(x[:,0],x[:,1],c=y,s=50,cmap='plasma')

<matplotlib.collections.PathCollection object at 0x04E1BBF0>

>>> plt.show()
```

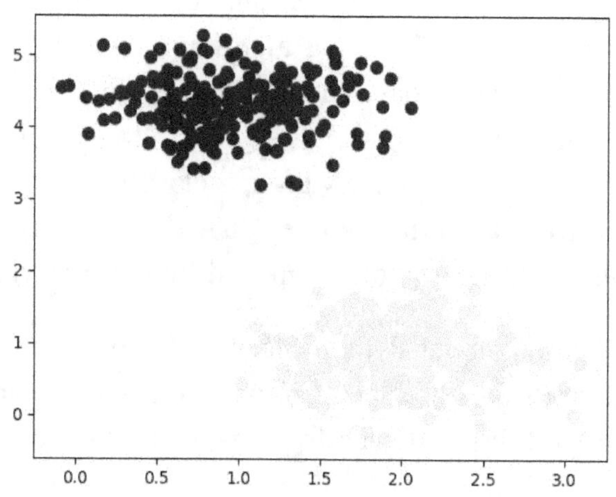

Machine Learning Algorithms in Python – SVM

```
>>> import numpy as np

>>> xfit=np.linspace(-1,3.5)

>>> plt.scatter(X[:, 0], X[:, 1], c=Y, s=50, cmap='plasma')

<matplotlib.collections.PathCollection object at 0x07318C90>

>>> for m, b, d in [(1, 0.65, 0.33), (0.5, 1.6, 0.55), (-0.2, 2.9, 0.2)]:

yfit = m * xfit + b

plt.plot(xfit, yfit, '-k')

plt.fill_between(xfit, yfit - d, yfit + d, edgecolor='none',
```

color='#AFFEDC', alpha=0.4)

[<matplotlib.lines.Line2D object at 0x07318FF0>]

<matplotlib.collections.PolyCollection object at 0x073242D0>

[<matplotlib.lines.Line2D object at 0x07318B70>]

<matplotlib.collections.PolyCollection object at 0x073246F0>

[<matplotlib.lines.Line2D object at 0x07324370>]

<matplotlib.collections.PolyCollection object at 0x07324B30>

```
>>> plt.xlim(-1,3.5)
```

(-1, 3.5)

```
>>> plt.show()
```

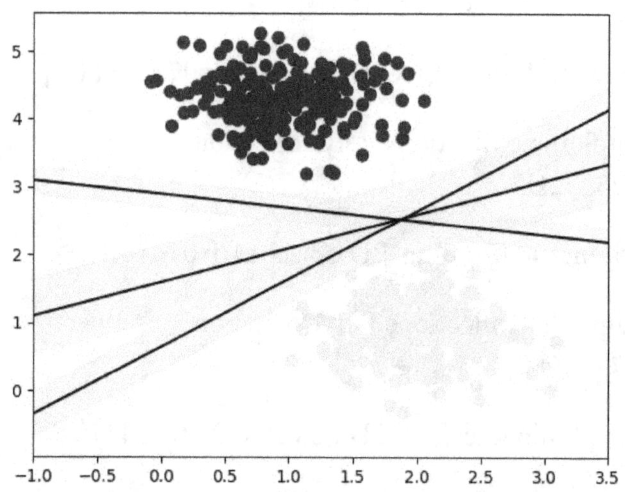

Machine Learning Algorithms in Python – Support Vector Machine

e. Bayes Naïves

Naive Bayes is a method of classification based on the theorem given by Bayes. This takes for granted independence between predictors. A classifier for Naive Bayes will presume a function in one class is unrelated to the other. Take a fruit. If it's round, red, and 2.5 inches in diameter, this is an apple. A Naive Bayes classifier would claim these features lead independently to the probability that the fruit is an apple. This is even though the features are mutually related.

Construction of a Naive Bayesian model is simple for very large data sets. The model is not only very basic,

```python
>>> from sklearn.naive_bayes import GaussianNB

>>> from sklearn.naive_bayes import MultinomialNB

>>> from sklearn import datasets

>>> from sklearn.metrics import confusion_matrix

>>> from sklearn.model_selection import train_test_split

>>> iris=datasets.load_iris()

>>> x=iris.data

>>> y=iris.target

>>>
x_train,x_test,y_train,y_test=train_test_split(x,y,test_si
ze=0.3,random_state=0)

>>> gnb=GaussianNB()

>>> mnb=MultinomialNB()

>>>
y_pred_gnb=gnb.fit(x_train,y_train).predict(x_test)

>>> cnf_matrix_gnb    =    confusion_matrix(y_test,
y_pred_gnb)

>>> cnf_matrix_gnb
array([[16, 0, 0],
 [ 0, 18, 0],
```

[0, 0, 11]], dtype=int64)

>>>   y_pred_mnb   =   mnb.**fit**(x_train, y_train).**predict**(x_test)

>>>   cnf_matrix_mnb   =   **confusion_matrix**(y_test, y_pred_mnb)

>>> cnf_matrix_mnb

array([[16, 0, 0],

[0, 0, 18],

[0, 0, 11]], dtype=int64)

F. kNN (k-Nearest Neighbors)

This is a classification and regression Python Machine Learning algorithms-mainly for classification. This is a supervised learning algorithm that considers different centroids and compares distance using a usually euclidean function. Then it analyzes the results and classifies each point into the group to optimize it to be placed with all the points closest to it. It classifies new cases using its neighbors majority vote k. The case which it assigns to a class is the most common among its closest neighbors to K. Uses a distance function for this.

i. Testing and Training   the entire dataset

>>> from sklearn. datasets import load_iris

>>> iris=**load_iris**()

```
>>> x=iris.data

>>> y=iris.target

>>> from sklearn.linear_model import LogisticRegression

>>> logreg=LogisticRegression()

>>> logreg.fit(x,y)

LogisticRegression(C=1.0, class_weight=None, dual=False, fit_intercept=True,
intercept_scaling=1, max_iter=100, multi_class='ovr', n_jobs=1,
penalty='l2', random_state=None, solver='liblinear', tol=0.0001,
verbose=0, warm_start=False)

>>> logreg.predict(x)
array([0, 0, 0, 0, 0, 0, 0, 0, 0, 0, 0, 0, 0, 0, 0, 0, 0, 0, 0, 0,
0,                                                                0,
0, 0, 0, 0, 0, 0, 0, 0, 0, 0, 0, 0, 0, 0, 0, 0, 0, 0, 0, 0, 0, 0,
0, 0, 0, 0, 0, 0, 1, 1, 1, 1, 1, 1, 1, 1, 1, 1, 1, 1, 1, 1, 1, 1,
2, 1, 1, 1, 2, 1, 1, 1, 1, 1, 1, 1, 1, 1, 1, 1, 1, 2, 2, 2, 1, 1,
1, 1, 1, 1, 1, 1, 1, 1, 1, 1, 2, 2, 2, 2, 2, 2, 2, 2, 2, 2,
2, 2, 2, 2, 2, 2, 2, 2, 2, 2, 2, 2, 2, 2, 2, 2, 2, 2, 1, 2, 2,
2, 2, 2, 2, 2, 2, 2, 2, 2, 2, 2, 2, 2, 2, 2])

>>> y_pred=logreg.predict(x)

>>> len(y_pred)
```

150

```
>>> from sklearn import metrics

>>> metrics.accuracy_score(y,y_pred)

0.96

>>> from sklearn.neighbors import KNeighborsClassifier

>>> knn=KNeighborsClassifier(n_neighbors=5)

>>> knn.fit(x,y)

KNeighborsClassifier(algorithm='auto', leaf_size=30, metric='minkowski',
metric_params=None, n_jobs=1, n_neighbors=5, p=2, weights='uniform')

>>> y_pred=knn.predict(x)

>>> metrics.accuracy_score(y,y_pred)

0.9666666666666667

>>> knn=KNeighborsClassifier(n_neighbors=1)

>>> knn.fit(x,y)

KNeighborsClassifier(algorithm='auto', leaf_size=30, metric='minkowski',
metric_params=None, n_jobs=1, n_neighbors=1, p=2, weights='uniform')
```

```
>>> y_pred=knn.predict(x)
>>> metrics.accuracy_score(y,y_pred)
1.0
```

ii. Splitting into train/test

```
>>> x.shape
(150, 4)
>>> y.shape
(150,)
>>> from sklearn.cross_validation import train_test_split
>>> x.shape
(150, 4)
>>> y.shape
(150,)
>>> from sklearn.cross_validation import train_test_split
>>>
x_train,x_test,y_train,y_test=train_test_split(x,y,test_si
ze=0.4,random_state=4)
>>> x_train.shape
(90, 4)
```

```
>>> x_test.shape
(60, 4)
>>> y_train.shape
(90,)
>>> y_test.shape
(60,)
>>> logreg=LogisticRegression()
>>> logreg.fit(x_train,y_train)
>>> y_pred=knn.predict(x_test)
>>> metrics.accuracy_score(y_test,y_pred)
0.9666666666666667
>>> knn=KNeighborsClassifier(n_neighbors=5)
>>> knn.fit(x_train,y_train)
KNeighborsClassifier(algorithm='auto',    leaf_size=30,
metric='minkowski',
metric_params=None, n_jobs=1, n_neighbors=5, p=2,
weights='uniform')
>>> y_pred=knn.predict(x_test)
>>> metrics.accuracy_score(y_test,y_pred)
0.9666666666666667
```

```
>>> k_range=range(1,26)

>>> scores=[]

>>> for k in k_range:

knn = KNeighborsClassifier(n_neighbors=k)

knn.fit(x_train, y_train)

y_pred = knn.predict(x_test)

scores.append(metrics.accuracy_score(y_test, y_pred))

>>> scores

[0.95, 0.95, 0.9666666666666667,
0.9666666666666667, 0.9666666666666667,
0.9833333333333333, 0.9833333333333333,
0.9833333333333333, 0.9833333333333333,
0.9833333333333333, 0.9833333333333333,
0.9833333333333333, 0.9833333333333333,
0.9833333333333333, 0.9833333333333333,
0.9833333333333333, 0.9833333333333333,
0.9666666666666667, 0.9833333333333333,
0.9666666666666667, 0.9666666666666667,
0.9666666666666667, 0.9666666666666667, 0.95, 0.95]

>>> import matplotlib.pyplot as plt

>>> plt.plot(k_range,scores)

[<matplotlib.lines.Line2D object at 0x05FDECD0>]
```

>>> plt.**xlabel**('k for kNN')

Text(0.5,0,'k for kNN')

>>> plt.**ylabel**('Testing Accuracy')

Text(0,0.5,'Testing Accuracy')

>>> plt.**show**()

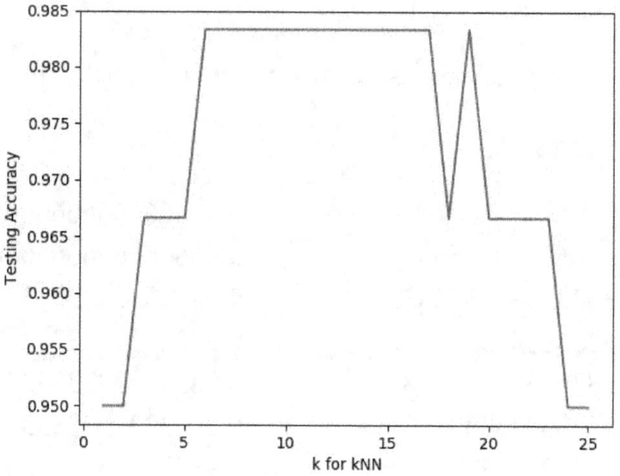

Machine Learning Algorithms in Python – k-Nearest Neighbors

G. k-Means

K-Means is an unsupervised algorithm which solves the clustering problem. It uses a number of clusters to classify the data. Within a class the data points are heterogeneous and homogeneous to peer groups.

```python
>>> import numpy as np

>>> import matplotlib.pyplot as plt

>>> from matplotlib import style

>>> style.use('ggplot')

>>> from sklearn.cluster import KMeans

>>> x=[1,5,1.5,8,1,9]

>>> y=[2,8,1.7,6,0.2,12]

>>> plt.scatter(x,y)
<matplotlib.collections.PathCollection object at 0x0642AF30>

>>> x=np.array([[1,2],[5,8],[1.5,1.8],[8,8],[1,0.6],[9,11]])

>>> kmeans=KMeans(n_clusters=2)

>>> kmeans.fit(x)
KMeans(algorithm='auto', copy_x=True, init='k-means++', max_iter=300,
    n_clusters=2, n_init=10, n_jobs=1, precompute_distances='auto',
    random_state=None, tol=0.0001, verbose=0)

>>> centroids=kmeans.cluster_centers_
```

```
>>> labels=kmeans.labels_
>>> centroids
array([[1.16666667, 1.46666667],
[7.33333333, 9. ]])
>>> labels
array([0, 1, 0, 1, 0, 1])
>>> colors=['g.','r.','c.','y.']
>>> for i in range(len(x)):
print(x[i],labels[i])
plt.plot(x[i][0],x[i][1],colors[labels[i]],markersize=10)
[1. 2.] 0
[<matplotlib.lines.Line2D object at 0x0642AE10>]
[5. 8.] 1
[<matplotlib.lines.Line2D object at 0x06438930>]
[1.5 1.8] 0
[<matplotlib.lines.Line2D object at 0x06438BF0>]
[8. 8.] 1
[<matplotlib.lines.Line2D object at 0x06438EB0>]
```

[1. 0.6] 0

[<matplotlib.lines.Line2D object at 0x06438FB0>]

[9. 11.] 1

[<matplotlib.lines.Line2D object at 0x043B1410>]

>>>
plt.**scatter**(centroids[:,0],centroids[:,1],marker='x',s=150
,linewidths=5,zorder=10)

<matplotlib.collections.PathCollection object at 0x043B14D0>

>>> plt.**show**()

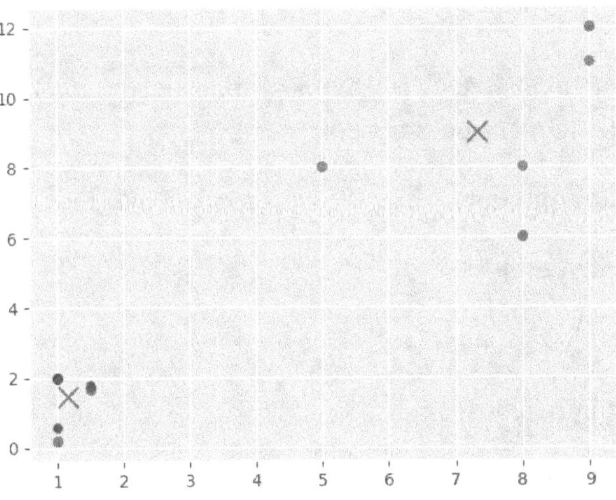

Machine Learning Algorithms in Python – K-Means

H. Random Forest

A random forest is an ensemble of decision trees. In order to classify every new object based on its attributes, trees vote for class- each tree provides a classification. In the forest takes the classification with the most votes.

```
>>> import numpy as np

>>> import pylab as pl

>>> x=np.random.uniform(1,100,1000)

>>> y=np.log(x)+np.random.normal(0,.3,1000)

>>> pl.scatter(x,y,s=1,label='log(x) with noise')
<matplotlib.collections.PathCollection object at 0x0434EC50>

>>> pl.plot(np.arange(1,100),np.log(np.arange(1,100)),c='b',label='log(x) true function')
[<matplotlib.lines.Line2D object at 0x0434EB30>]

>>> pl.xlabel('x')
Text(0.5,0,'x')

>>> pl.ylabel('f(x)=log(x)')
Text(0,0.5,'f(x)=log(x)')

>>> pl.legend(loc='best')
<matplotlib.legend.Legend object at 0x04386450>
```

>>> pl.**title**('A basic log function')

Text(0.5,1,'A basic log function')

>>> pl.**show**()

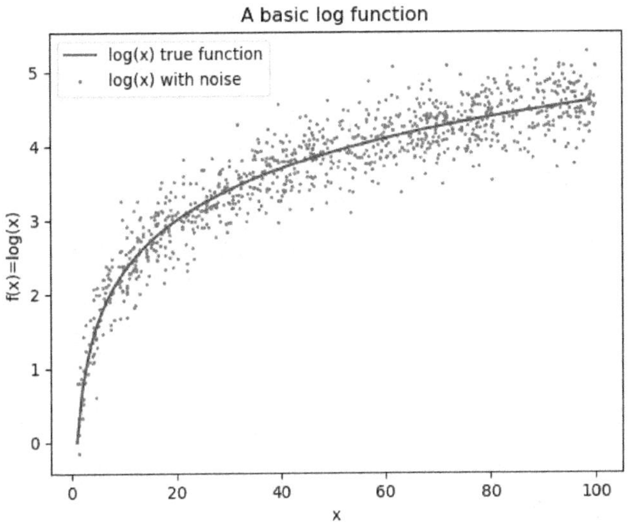

Machine Learning Algorithms in Python – Random Forest

>>> from sklearn.datasets import load_iris

>>> from sklearn.ensemble import RandomForestClassifier

>>> import pandas as pd

>>> import numpy as np

>>> iris=**load_iris**()

```
>>>
df=pd.DataFrame(iris.data,columns=iris.feature_names
)

>>> df['is_train']=np.random.uniform(0,1,len(df))<=.75

>>>
df['species']=pd.Categorical.from_codes(iris.target,iris.t
arget_names)

>>> df.head()

sepal length (cm) sepal width (cm) ... is_train species
0               5.1             3.5 ...            True  setosa
1               4.9             3.0 ...            True  setosa
2               4.7             3.2 ...            True  setosa
3               4.6             3.1 ...            True  setosa
4               5.0             3.6 ...            False setosa
[5 rows x 6 columns]

>>>
train,test=df[df['is_train']==True],df[df['is_train']==Fals
e]

>>> features=df.columns[:4]

>>> clf=RandomForestClassifier(n_jobs=2)

>>> y,_=pd.factorize(train['species'])

>>> clf.fit(train[features],y)
```

```
RandomForestClassifier(bootstrap=True,
class_weight=None, criterion='gini',

max_depth=None,                    max_features='auto',
max_leaf_nodes=None,

min_impurity_decrease=0.0, min_impurity_split=None,

min_samples_leaf=1, min_samples_split=2,

min_weight_fraction_leaf=0.0,           n_estimators=10,
n_jobs=2,

oob_score=False, random_state=None, verbose=0,

warm_start=False)
>>> preds=iris.target_names[clf.predict(test[features])]

>>>
pd.crosstab(test['species'],preds,rownames=['actual'],col
names=['preds'])
```

preds	setosa	versicolor	virginica
actual			
setosa	12	0	0
versicolor	0	17	2
virginica	0	1	15

So, in this Python Guide chapter, it was all about Machine Learning Algorithms. I hope you like some of our explanations.

CHAPTER 9

11 BEGINNER TIPS FOR LEARNING PYTHON PROGRAMMING

1: Code Daily

Consistency is extremely necessary when studying a new language. We recommend making a regular dedication to coding. It may be hard to believe, but programming plays a large part in muscle memory. Committing to coding daily should help your muscle memory grow. Though at first, it may seem daunting, consider starting small every day with 25 minutes and working your way up from there.

2: Write it Down

As you advance on your journey as a new programmer, you may wonder if not to take notes. Sure. You have to! Research suggests that it is most beneficial to take notes by hand for long-term retention. This will be particularly beneficial for those working towards the goal of becoming a full-time developer since many interviews will involve writing code on a whiteboard.

When you get started working on small projects and applications, writing will also help you prepare your code before heading to your computer. You will save enough time by figuring out what roles and classes you will need and how they communicate.

3: Go in interactively!

If you're learning for the first time about simple Python data structures (strings, lists, dictionaries, etc.) or debugging an application, the interactive Python shell will be one of your best learning instruments.

Ensure Python is downloaded and installed on your computer to use the interactive Python shell (also sometimes called a "Python REPL"). To help you do that, we have a step-by-step tutorial. Open your terminal and run python or python3 based on your installation, to enable the interactive Python shell. More specific directions can be found here.

4: Catch Breaks

It is important to step away when you're learning and absorb the concepts. The Pomodoro technique is commonly used and can help: you are working for 25 minutes, taking a brief break, and repeating the cycle afterward.

Taking breaks is crucial to having an effective study session, particularly when a lot of new knowledge is being taken in.

Breaks are especially necessary when debugging. If you are hitting a bug and can't figure out exactly what's going wrong, take a break. Leave your computer, go on a walk or chat with a friend.

In programming, the code must exactly obey the laws of a language and logic, so even lacking a quotation mark breaks everything. Fresh eyes make a considerable difference.

5: Turn into a Bug Bounty Hunter

Speaking of hitting a bug, it's inevitable you'll run into bugs in your code once you start writing complex programs. It happens to us all! Don't let you get irritated by bugs. Rather, welcome these moments of confidence and fancy yourself a bug bounty hunter.

It 's critical to have a systematic approach when debugging to help you find where things are breaking down. Going through the code in the order it is executed and making sure that every component works is a perfect way to do it.

If you've got an idea of where things could break down, insert and run the following line of code into your script import PDB; PDB.set trace). This is the debugger for Python and will drop you into interactive mode. The debugger can also run python -m PDB < my file.py > from the command line.

Make It Collaborative

Once things start to stick, work together to expedite your learning. Get the most out of working with others. That will help you with the few strategies below.

6: Surround yourself with those who learn.

Although coding may seem like a solitary activity, when you work together, it works the best. When you learn to code in Python, you must surround yourself with other people who are also learning. This will let you share the tips and tricks you 're learning along the way.

Don't worry, unless you know anybody. There are many ways to meet with other people who are passionate about learning Python! Find meetups or local events or enter PythonistaCafe, a peer-to-peer learning group for Python enthusiasts like you!

7: Teaching

It's said teaching something is the best way to learn something. If you are learning Python, this is valid. There are numerous ways to do this: whiteboarding with other Python lovers, writing blog posts explaining newly learned concepts, recording videos explaining something you've learned, or simply talking to yourself on your computer. Each of these strategies will both strengthen your understanding and expose any gaps in your understanding.

8: Pair programming

Pair programming is a strategy involving two developers collaborating to complete a job at one workstation. The two developers switch between being the "navigator" and the "driver." The code is written by the "driver," whereas the "navigator" helps guide the problem solving and reviews the code as written. Turn often to allow both sides to benefit.

Pair programming has many advantages: it allows you not only to have somebody review your code but also to see how somebody else might think about a problem. When you get back to coding on your own, being exposed to different ideas and ways of thinking will help you solve the problems.

9: Ask "GOOD" Questions

People always say that there is no such thing as a bad question, but it is possible to ask a bad question when it comes to programming. When you ask someone who has little or no context on the problem you are trying to solve for help, it's best to ask GOOD questions following this acronym:

G: Give context to what you're trying to do, and describe the problem clearly.

O: Set out the stuff that you've already tried to repair.

O: Offer your best guess of what could be the problem. This helps the person who helps you not only know what you're thinking about but also know you've made some thinking on your own.

D: Example of what's going on. Include the code an explanation of the steps you were executing, which led to the error, and a traceback error message. This way, the person assisting does not need to try to recreate the problem.

Good questions can be time-saving. Skipping each of these measures can lead to conflict-causing back-and-forth conversations. As a beginner, you want to ensure that you ask good questions so that you practice sharing your thought process so that people who support you can continue to support.

Make a point

Most, if not all, Python developers to whom you're speaking will tell you that to learn Python, you need to learn by doing so. Exercising can only take you so far: by building up, you learn the most.

10: Build something, anything, everything

There are lots of small exercises for beginners that will help you become confident with Python, as well as develop the muscle memory we discussed above. Once you have a solid grasp of basic data structures (strings, lists, dictionaries, sets), object-oriented programming, and class writing, it's time to start building!

What you're building isn't as critical as how you create it. The building journey is what will teach you the most. You can only learn so much from courses about Real Python. Most of your learning should come from creating something using Python. You'll be taught a lot about the problems you will solve.

11: Add to Open Source

Software source code is freely accessible through the open-source model, so everyone can collaborate. There are plenty of Python libraries that are open-source projects and contribute. Also, several companies publish open-source projects. This means that the engineers working in these companies can work with code written and produced.

Contributing to an open-source Python project is a perfect way to build meaningful learning opportunities. Let's say you want to apply for a bug fix: you send a "pull request" to patch your fix into the application.

The project managers will then review your work and give comments and suggestions. This will encourage you to learn Python programming best practices, as well as interact with other developers.

CHAPTER 10

CHOOSING A PROJECT PLATFORM

You need to develop your program to run on a platform, so you can use your program for people who lack other technical knowledge. The desktop, web, and command-line are the 3 major platforms for which you would like to build your projects.

Website

Web applications are applications running on the web; they can be accessed without downloading on any device, providing access to the internet is available. If you want someone with internet access to access your projects, it needs to be a web application.

A Web application has a front end and a back end. The back-end is where the business logic is: the data is manipulated and stored by your backend code. The front end is the application interface: your front end code will determine what a web application looks like.

Your main focus as an intermediate Python developer will be on the back-end code. However, the front-end code is also essential, so you'll need some CSS, HTML, and maybe JavaScript knowledge to construct a simple-looking GUI. Just the basics would suffice.

Another option is to use Python for the front end as well as back end. You can focus on Python code alone thanks to the anvil library, which eliminates the need for HTML, JavaScript, and CSS.

With Python, you can create web applications through web frameworks such as Django and flask. The list of frameworks used to build web apps using Python is long. There's plenty to choose from, but the most popular web frameworks remain Django and flask.

Desktop Graphics

It is through an application whenever you perform a task on your PC, whether it's a desktop or laptop. You can also make your desktop applications as an intermediate Python developer.

As you've seen with web applications, you don't have to learn any front-end technology to create your Graphical

User Interface (GUI) applications. All the parts can be built using Python.

Frameworks exist to build your desktop applications. PySimpleGUI is one of them, and an intermediate Python developer finds it pretty user-friendly.

An advanced GUI framework such as PyQt5 is quite powerful, but perhaps it has a steep learning curve.

The Desktop GUI program you build will operate on all of the Windows, Linux, or Mac operating systems. After creating the project, all you need to do is compile it to an executable for your preferred operating system.

Control-Line

Command-line apps are those apps that work in a console window. This is the Windows command prompt, and the Linux and Mac Screen.

You would click to use a web or GUI application, but for command-line applications, you would type in commands. Command-line software users need to have some technical expertise because they may need to use commands.

Command-line applications can not be as beautiful or as user-friendly as web or GUI applications, but this does not make them less effective than web or GUI.

By adding colors to the text, you can enhance the look of your Command-line applications. There are libraries that you can use to color, such as color and Colorama. You should spice up things and use a few colors.

You can build your applications using frameworks such as Docopt, Argparse, and click.

Ideas for web project

You'll see project ideas for the Web in this section. These ideas for a project can be classified as tools of utility and education.

Here are the ideas behind the project:

- Content Aggregator
- Post -It Note
- Quiz Application
- URL Shortcutter
- Regex Request Device

Aggregating Content

Content is gold. From blogs to social media platforms, it exists everywhere on the web. To keep up, you need to continuously check the Internet for new information. One way of staying updated is to manually check all of the sites to see what the new posts are. But it's time-consuming, and it's quite tiring.

This is where the aggregator of content comes in: A aggregator of content collects information from different places online and collects all of that information in one place. So, to get the latest information, you don't have to visit multiple sites: one site is enough.

With the content aggregator, you can get all the latest information from one site, which aggregates all the content.

Examples of Content Aggregators

Here are some ideas for the Content Aggregator implementations:

• Hvper

• AllTop

Application Info

The main goal of this idea for a project is to aggregate content. Next, you need to know from which places you 're going to use the web aggregator to get the web. You can then use libraries such as requests to submit HTTP requests and BeautifulSoup to decode and scrape the content from the pages that are required.

As a background process, your application can implement its content aggregation. Therefore libraries like ap-scheduler or celery can help. You can have an ap-scheduler try. It is great for small processes in the background.

You'll need to save it somewhere after you scrap content from different sites. So, you will use a database to save the content that has been scraped.

Extra Challenge

You can add more websites to an even harder challenge. This will help you learn how websites research and extract knowledge.

You can also have users subscribing to certain sites you are aggregate. The content aggregator will then, at the end of the day, send the articles for that day to the user's email address.

Regex Call Tool

You and I handle text daily. There is a structure to this book, which is also text. This makes understanding easier for you. Sometimes, you need to find some information in text, and it may be ineffective to use the regular search tool in text editors.

It's here where the Regex Query Tool comes in. A regex is a collection of strings, which means that the regex query tool can test the validity of the queries. Once the regex matches patterns in the text, this tells the user and highlights the patterns that match. Your Regex Query Tool will check the user's passed in regex strings for validity.

Users can easily check the validity of their regex strings over the web with the Regex Query Tool. This makes it easier for them, rather than having to use a text editor to check the strings.

For instance, Regex Query Tools

Here are several Regex Query Device Implementations:

• RegexTester

• FreeFormatter

Application Info

The main aim of this type of project is to tell the user the validity of the query strings that are entered. You can make it give a negative or positive response like Query String Is True and Query String Is Invalid, which implements the positive answer in green and the negative in red.

The query method doesn't need to be built from scratch. You can use the standard re library of Python, which you can use to run the query strings on the text you are entering. When the query string matches nothing, the re library returns None, and it returns the matched strings if positive.

Some users may not fully understand regex, so you can do a page explaining how regex works. You can create documentation that is sufficiently interesting to keep users excited about learning and understanding regex.

Extra Challenge

It's fine to do a project that just returns the regex validity. But you can add a replacement function, too. This means that the application checks for regex validity and also allows users to replace the matched strings with something else. So the tool is no longer a tool for finding but also a tool for replacing.

URL Cutter

URLs can be incredibly long and not easy to use. It's difficult when people share links or even try to remember a URL because most URLs are filled with difficult characters and don't make meaningful words.

Here's where the Shortener URL comes in. A URL Shortener decreases the characters or letters in a URL and makes it easier to read and remember. A URL close to xyz.com/wwryb78&svnhkn%sghq? Sfiyh may be shortened to xyz.com/piojwr.

URLs are a joy to deal around with with with the URL Shortener.

For example, URL Shorteners

Here are some URL Shortener Implementations:

• Bitter

• MeShort

Tech Details

This project idea's main aim is to shorten the URLs. The main task the application will perform is to shorten URLs, and then redirect users to the original URL when visiting the shortened URL.

The users input the original URL in the application, and the result will be the new, shortened URL. To do so, a blend of random and string modules can be used to generate the characters for the shortened URL.

You will need to save the original and shortened URLs in a database as users visit the shortened URL days, months, or years later. When a request enters, the application checks whether the URL exists and redirects to the original, or redirects it to a 404 page.

Extra Challenge

Creating a shortened URL with random characters makes them shorter, random URLs better than the long ones. But, you can make the users' outcomes better. To customize URLs, you can add a feature, so users can customize the created URLs themselves.

Unquestionably, a custom xyz.com/mysite URL is

better than a randomly produced xyz.com/piojwr URL.

Post-Remark

Having many thoughts and ideas in a day is normal, but forgetting is also normal. One way of working around missing things is to pin them down before they disappear into thin air. Although some of the thoughts and ideas that were forgotten may be insignificant, some may be very important.

That is where a Post-It note comes in: A Post-It note is a small paper with low-tack adhesive on the back, making it fixable to surfaces such as papers, walls. Post-it notes make things easier to pin down. The idea for the Post-It Note project is something similar. It enables users to pinpoint things down, making them accessible anywhere, as it is a web application.

With the Post-It note, people can now track items everywhere, without the fear of missing stuff or misplacing notes — which is difficult for paper notes.

Post-it Notes for examples

Here are some Post-It Note Implementations:

• Note.ly

• Pinup

Tech Details

This project's main objective is to allow users to track thoughts. This implies that each user will have their notes, so they will need to have an account creation feature in the application. This ensures every user's notes remain private to them.

Django comes with an authentication system for users, so it might be a good choice. You can use other frameworks such as a bottle or flask, but the user authentication system will have to be implemented on your own.

Since users will need to classify their notes under different parts, the application would be more useful if a function is introduced to allow users to categorize their notes.

You may need notes on algorithms and data structures as an example, so you'll need to be able to distinguish the notes in those categories.

You may need to store every user's details and notes so that a database is an integral part of this project. The MySQLdb module can be used for a PostgreSQL database if you wish to use a MySQL database or a psycopg2 module. You can use other modules, but it all depends on the database you choose to use.

Extra Challenge

Since users are human in forgetting their ideas, it is also human for them to forget that they even made a note somewhere. You can add the functionality to remind users of their notes. This feature allows users to set a notification time, and the application can send the alert by email to the users when it's time.

Quiz Application

Know-how is strength. There are so many things to learn in the world, and quizzes help to test the understanding of those concepts. You do not have to understand everything about the language, as an intermediate Python developer. One way to figure out things you don't fully understand is to take tests.

That is where the question for a Quiz comes in. The Quiz Program will be asking users questions and requesting the correct answers to those questions. Think of the application of the Quiz as some kind of questionnaire.

Special users can call administrators will be allowed to create tests with the Quiz Application, so regular users can test their understanding and answer the questions of the subjects in the quiz.

Types of Quiz requests

Here are a few implementations of the concept for the Quiz Application:

• myQuiz

• Kahoot

Application Info

This project's main aim is to set quizzes and get people to answer them. Users should, therefore, be able to put questions, and other users should be able to answer those questions. Afterward, the application will show the final score and the correct answers.

If you want users to be able to record their ratings, you may need to incorporate an account creation feature.

Users creating the tests should be able to create tests by simply uploading a text file with the questions and answers. The text file will have a format to choose from so that the application can convert from a file to a quiz.

To do this project, you will need to implement a database. For each user, the database will store the questions, potential answers, correct answers, and the scores.

Extra Challenge

You can allow the users to add timers to the quizzes for more of a challenge. In this way, quiz developers will decide how many seconds or minutes a user will spend in the quiz on any question.

Having a quiz-sharing feature will be awesome, too, so users can post fun quizzes with their friends on other platforms.

Ideas for GUI project

You will see project ideas for Graphical User Interfaces in this section. These ideas for a project can be classified as tools for entertainment, finance, and utility.

Here are the ideas behind the project:

• MP3 Player

• Alarm Tool

• Film Administrator

• Expenditures Tracker

MP3 Player

Audio is as important as, if not more important than, text today. Since audio files are digital files, you will need a playable tool to play them. You will never be able to listen to the content of an audio file without a player.

This is where the Player for MP3 enters. The MP3 Player is a computer where MP3s and other digital audio files are playable. This idea for a project MP3 Player GUI is trying to emulate the physical MP3 Player. You can create applications that let you play MP3 files on your laptop or desktop.

Upon construction of the MP3 Player project, users can play their MP3 files and other digital audio files without having to buy a physical MP3 Player. They'll be able to use their computers to play the MP3 files.

MP3 Player Tests

Here are some MP3 Player Implementations:

The MusicBee

Foobar2000

Tech Details

This project's main aim is to allow users to play both digital audio files and MP3. The application needs to have a simple but beautiful user interface to be engaging to users.

You can have an interface with which to list the available MP3 files. You can also give users the option of listing other, non-MP3 digital audio files.

The users will also expect the MP3 Player to have an interface that will display information about the file being played. Some of the information that you can include are in minutes and seconds the file name, its length, the amount not played, and the amount played.

Python has libraries that can play audio files in a few lines of code, such as Pygame, which enables you to work with multimedia files. You can also check out Simple Audio and Pymedia.

Many digital audio files can be handled by these libraries. They can handle certain types of files, not just those of MP3 files.

You can also introduce a feature that allows users to build a playlist. To do this, you will need a database to store information on the playlists that were made. The sqlite3 module from Python lets you use the SQLite database.

In this case, the SQLite database is a better option, since it is more file-based and simple to set up than other SQL databases. While SQLite is file-based, data conservation is better than a regular file.

Extra Challenge

You may add a function to allow the MP3 player to repeat current playing files, or shuffle the list of files to be played, for a more exciting challenge.

A function that allows users to increase and decrease the play speed of the audio file may also be introduced. This will be useful to users because they will be able to play files at a faster or slower speed than normal.

Tool to alarm

As they say, "Tide and time wait for no man." But with so many things happening in our lives, it's hard not to lose track of time. A recall is required to be able to keep track of the time.

This is where the Alarm Tool comes in. An alarm is a device giving an audio or visual signal about a particular condition. This idea for the project Alarm Tool is an attempt to create an alarm as software. When a certain condition is met, the Alarm Tool gives an audio signal. In this case, the set time is the stated state.

With the Alarm Tool, users may set alarms at certain times of the day to remind them of things. The project Alarm Tool can operate from the user's laptop or desktop computer, so they don't have to buy a physical timer.

Alarm Tools examples

Here are several Security Device Implementations:

• TimerForMac

• FreeAlarmClock

Tech Details

This project's main objective is to trigger the audio signals at certain times of the day. So, the most important parts of the Alarm Tool are timing and the audio signal to be played.

The Alarm Tool should allow alarm formation, editing, and deletion by users. It should also have an interface that lists all the alarms if the user has not deleted them. So, it will mention alarms that are active and inactive.

The application must play tones at the set time since it is an alarm. There are audio-playing libraries, such as the Pygame library.

The application has to continue checking for set alarm times in your code logic. When the time is reached, the playing of the alarm tone activates a feature.

As the application must search for fixed alarm times, this means that the application will save the alarms in a database. The database will store such items as date, time, and location of the alarm.

Extra Challenge

You can also allow users to set recurring alarms as an extra feature. They will be able to set alarms on certain days of the week, each week, that will ring at some time. For instance, an alarm can be set every Monday at 2:00 PM.

You can add a snooze feature, so your users can snooze alarms rather than just dismiss them.

File Manager

The number of files on an average PC user's personal computer is fairly high. If all of those files are placed in a single directory, navigating and finding files or directories would be difficult. So, the files need to be arranged and managed correctly.

This is where a director of files comes in. A file manager allows users to access files and folders through a user interface. While files can be controlled via the command-line, not all users know how to do that.

Users can arrange, access, and manage their files and directories properly with a file manager, without knowing how to use the command line. Some of the tasks that a file manager does include copying, moving, and renaming files or directories.

Examples of Tools in File Manager

Here are some ideas for the File Manager implementations:

• FreeCommander

• Explorer+++

Tech Details

The file manager project's main objective is to provide the users with an interface to manage their files. Users want a file manager with a good looking and easy to use file management tool.

You can use the PySimpleGUI library without having to deal with a lot of complexity to create unique user interfaces with a powerful widget.

Your users should perform simple tasks such as creating new directories or emptying text files. They should also be able to copy directories or files and move them.

For this project, the sys, os, and Shutil libraries will be very useful, because they can be used to perform actions on the context files while the user clicks away.

Today, grid views and list views are common views, so you can implement both in the app. This gives the user the option to select which view option fits them.

Extra Challenge

To make the file manager somewhat more advanced, a search function can be implemented. In this way, users can search for files and directories without having to manually find them.

Also, you can implement a sorting feature. This will allow users to sort files by order, such as time, alphabetical order, or size.

Expenditure Tracker

We have daily expenses ranging from foodstuffs to clothing to bills. There are so many expenses that it is normal to lose track of them and continue to spend until we are nearly out of cash. A tracker will help people look out for their expenses.

It's here that the expense tracker comes in. An expense tracker is a software tool enabling users to keep track of their spending. Depending on how advanced it is, it can also analyze the expenses, but let's keep it simple for now.

Users can set a budget with the expense tracker and track their expenditures so they can make better financial decisions.

Examples of Tracker Expenses

Here are some of the Cost Tracker Concept implementations:

• Gnucash

• Buddi

Tech Details

This project's main aim is to keep track of user expenses. Some statistical analysis must be done to allow users to get accurate information about their expenditures and to help them spend better.

While monitoring the expenses is crucial, it is also necessary to have a good GUI. You can create a unique interface with PySimpleGUI to improve the user's experience.

PyData libraries like matplotlib and pandas can help to create an expense tracker.

The pandas' library can be used in the analysis of data, and the matplotlib library can be used for graph plotting. Graphs will provide the users with a visual representation of their expenses, and usually, a visual representation is easier to understand.

The application will get users' data. The data here are the expenses inputted. So, the expenses will have to be stored in a database. For this project, the SQLite database is a good database option because it can be set up easily. The SQLite database can be used with the sqlite3 module.

Extra Challenge

For your users to benefit from this project, they will have to periodically enter their expenses which will slip their minds. Implementing a notification feature might be helpful for you. So at certain times of the day or the week, the application will send a notification, reminding them to use the expense tracker.

Ideas for a command-line project

In this section, you will see command-line project ideas. The ideas discussed for the project can be classified as utility tools.

Here are the ideas behind the project:

• Contact Book

• File Connectivity Checker

• Bulk File Rename Tool

• Directory Tree Generator

Contact library

We come across a lot of people every day. We make friends and acquaintances. We are getting their contacts to keep in contact later. Sadly, keeping the contact details, you have received can be difficult. One way to achieve this is to write down the contact details. But this is not secure as you can easily lose the physical book.

Here's where the project Contact Book comes in. A contact book is a tool to save details about a contact, such as name, telephone number, address, and email address. For this project contact sheet, you can create a software tool that lets people save and find contact information.

Users can save their contacts with the concept of the contact book project, with less chance of losing the saved contact information. It will always be accessible through the command-line from their computer.

Examples of Software for Contacts

There are Contact Book applications, but finding Contact Book products on the command line is rare, as most are Mobile, Web, or GUI applications.

Below are some Contact Book Implementations:

• Pobuca Connect

• Simple Contacts

Tech Details

This project's main purpose is to save the contact information. Set up the user's commands for entering contact details. You can use the argparse, or click frameworks on the command line. They abstract a lot of complicated things, so when executing commands, you just have to concentrate on the logic to be run.

Some of the features that you should implement include the contact deletion commands, listing saved contacts, and updating contact information. You may also allow users to list contacts using different criteria, such as alphabetic order or date of contact formation.

The SQLite database will be fine for saving contacts as it is a command-line project. SQLite is an easy-to-use setup. You can save the contact information in a file, but a file doesn't provide the advantages you can get from using SQLite, such as performance and protection.

The Python sqlite3 module will be very useful in using the SQLite database in this project.

Extra Challenge

Remember, how do you store the database on your computer? What if something, like the user losing their files, happens? It means that they will lose contact information, too.

You can also push yourself and transfer your account to an online storage site. To do so, at some times, you can upload database files to the cloud.

Also, you can add a command allowing users to backup the database themselves. This way, if the database file is lost, the user can still have access to the contacts.

You should remember that some form of identification may be needed so that the contact book can say which database file belongs to which user. Implementing an authentication function for users is one way to get things done.

Site Connectivity Checker

When you visit a URL, you expect your browser to view the requested links. But not always, this is the case. Sometimes the sites may be down so that you won't get the results you want. Instead, error notices will be sent to you. You can continue to try a down site until it arrives, and you get the information you need.

This is where the project joins the Site Connection Checker. The Site Connectivity Checker visits a URL and returns the URL status: it may or may not be active. At intervals, the Site Connectivity Checker will visit the URL, returning the results of each session.

Instead of visiting a URL manually, a Site Connectivity Checker will do all the manual work for you. So, you'll just get the results of the search without having to

waste time on the browser, waiting for the site to go live.

Site Connectivity Checker Examples

Here are some of the Site Connection Checker implementation ideas:

• Site24x7

• Ping

Application Info

This project's main objective is to check the status of the sites. So, to test a website's status, you must write code.

You may choose either to use TCP or ICMP for your connections. One for checking out is the socket module. Socket Programming can also be read in Python (Guide).

You can attach commands to allow users to add and remove sites from the list of sites to be reviewed through your chosen system, be it the button, Docopt, or Argparse frame.

Also, users should be able to start the tool, stop it, and determine the intervals.

Since the list of files to be checked will have to be saved, you can either use an SQLite database through the sqlite3 module or save it in a file (just a list of sites).

Extra Challenge

The application will monitor the site's connection status and show the results to the command-line. But that will allow the user to keep the command-line verified.

You will raise the difficulty and get a notification feature implemented. The notification feature may be a sound that is played in the background to alert the user when a site changes its status. To store a site's prior status, you'll need a database. That is the only way the tool can tell about the change of status.

Bulk File Rename Tool

Sometimes, all the files in a directory must be named according to certain conventions. For example, in a directory with File0001.jpg, you can name all the files, where the numbers are increasing based on the number of files in the directory. That can be stressful and repetitive to do manually.

The Bulk File Rename Tool enables users to rename a large number of files, without the need to rename files manually.

That saves a lot of time for the users. It saves them the trouble of having to do routine, boring work, and making mistakes. Users can rename files in a few seconds with the Bulk File Rename Tool without any mistakes.

Examples of Tools to Rename Bulk Files

Here are several implementations of the concept of renaming Bulk File:

• Rename

•   Ren

Application Info

The main goal of this idea for a project is to rename files. Therefore, the application has to find a way to manipulate the target files. The libraries os, sys, and shutil will be useful for much of this project.

Using naming conventions, your users will be able to rename all the files in the directory. They should, therefore, be able to pass choice in the naming convention. If you understand how regex functions, the regex module can help suit the appropriate naming patterns.

A user may want to pass as part of the commands in a naming convention like my files, and expect the tool to rename all the files like myfilesXYZ, where XYZ is a number. Also, they should be able to select the directory where the files to be renamed are.

Extra Challenge

The big challenge in this project is renaming all of the files in a directory. However, users may need to name just a specific number of files. To test your skills, you can implement a feature that allows users to select the number of files to be renamed, rather than all the files.

Note that renaming only several files allows the application to arrange the files based on alphabetical order, file size, or file creation time, depending on the user 's needs.

Tree Directory Generator

Directories are similar to family trees: each directory has a special relationship with other directories. No directories, except an empty root directory, ever remain on its own.

It is difficult to see the connection between folders when you're dealing with files and directories because you can only see what's in the current directory. You use a file manager, or you work from the command line.

You can see the relation between files and directories like a tree or map with a Directory Tree Generator.

This makes the positioning of the files and directories easier to understand. When explaining certain concepts a directory treemap is important, and a generator of directory tree

makes it easier to obtain a visual representation of the relationships between files and directories.

Examples of Tree Generators in Directories

Here are some Directory Tree Generator Implementations:

• Dirtreex

• Tree

Application Info

The Directory Tree Generator's main objective is to visualize the connections between files and directories. The os library can be very useful to list the files and directories in a selected directory.

Using a framework like Argparse or Docopt helps to abstract loads of things, allowing you to concentrate on writing code for the logic of the application.

In the logic of the application, you can decide how files or directories you want to represent. It's a brilliant way to go about it using different colors. The colored library can be used to print files and folders in various colors.

Also, you can determine how far you want the Directory Tree Generator to go. For instance, if a directory has children directories twelve deep levels, you may only decide to go as deep as the fifth level.

You can also let the user decide how deeply they want the Directory Tree Generator to go if you wish.

Extra Challenge

Since the results will be on the command-line of the generated directory tree, you can go one step further. You can have the generator create directory tree images, so it will essentially turn the text into an image.

The pillow library would be useful for doing this.

Tips to Project Works

It can be tough working on projects. That's one reason why project motivation and interest will make it a less daunting task.

If you're interested in a project, you 're going to be able to spend time researching as well as finding libraries and tools to help with the project.

Here are a few hints:

• Consider motivational sources

• Bridge the project to subtasks

• Do some subtask research

• Make every subtask, one step at a time

• Get help in case you 're stuck

• Put up the subtasks

CONCLUSION

Right now, Python's fun and stuff but note ... Python is the language of complete programming. Any code that you download from GitHub, the Python Package Index, or anywhere else can be malicious, and the firewalls of most people will almost certainly go unnoticed. Uploading your kit to PyPi doesn't take long for anyone to update, so you need to be patient and alert.

I believe the trial has shown that the use of Python as the main teaching language is both possible and desirable:

• It's Free (as in both cost and code source).

• Installing a Windows PC is trivial, enabling students to take further interest in it. The hurdle of installing a C compiler or a Pascal on Windows machine is either too costly or too complicated for many;

• It is a versatile tool that allows both conventional procedural programming and modern OOP teaching; it can be used to teach a wide variety of transferable skills;

• It is a reality programming language that can and will be used in academia and the business world;

• It tends to be simple to learn, and this, in combination with its many libraries, provides the possibility of easier student growth that makes the course more demanding and varied;

• And above all, its clean syntax gives students greater comprehension and enjoyment;

Python will be used as the language of instruction for the first year. If used, more programming and less of the peculiarities of a given language will be possible to teach students. It is inadvisable to teach a mid-level language like C in a single day. Too much time must be spent teaching C and not enough time for students with no programming experience to teach generic skills.

Python's use as the language of the first year is not a dead-end. I tried to emphasize that Python allows the teaching of programming concepts that are widely applicable. In no way does its use preclude the use of C in a more advanced course. In reality, students who go on to use C in later years would have a better background in concepts from their introduction to programming than they would have from an introduction focused on the C. I assume that if implemented using Python, more students will move on to advanced programming because implementing programming using C would frustrate and scare off a large number of learners. To conclude, Python provides the best possible balance between applicability and teachability.

9 798656 373784